The Unsettled Clock

The Unsettled Clock

The Persistence of Time Pluralism

Kevin Birth

berghahn
NEW YORK • OXFORD
www.berghahnbooks.com

First published in 2026 by
Berghahn Books
www.berghahnbooks.com

Library of Congress Cataloging-in-Publication Data
A C.I.P. cataloging record is available from the Library of Congress
Library of Congress Cataloging in Publication Control Number:
2026006063

British Library Cataloguing in Publication Data
A catalogue record for this book is available from the British Library

EU GPSR Authorized Representative
LOGOS EUROPE, 9 rue Nicolas Poussin, 17000, LA ROCHELLE, France
Email: Contact@logoseurope.eu

ISBN 978-1-83695-539-9 hardback
ISBN 978-1-83695-540-5 epub
ISBN 978-1-83695-541-2 web pdf

https://doi.org/10.3167/9781836955399

Contents

Illustrations

Preface

The central idea guiding this project has been: clocks could be different, and at some point, will probably have to be different from what they currently are in order to allow us to think about the diversity of rhythms and cycles in our universe.

This perspective emerged out of a confluence of several events that followed the publication of my book *Objects of Time*.[1] The first was my becoming an observer to the leap-second debate with the concern that was agreed upon by most of those involved that there should only be one reference timescale, even if they disagreed over what that timescale should be. The second was my being asked by the Frick Collection to give a lecture in 2014 associated with its special exhibit "Precision and Splendor," which displayed some of the exceptional clocks in its collection. That work on clocks displaying multiple hour systems led me to recognize that the current emphasis on a single timescale was a recent development. A third event was my being asked to present at a conference on the psychology of time perception, a field in which the understanding of human physiology is, in my opinion, constrained by the metaphor of the clock.

At the same time, I was growing frustrated by the challenges faced by ethnographic discussions of concepts of time. As I knew firsthand from my work in Trinidad, anthropology faced a challenge in documenting concepts of time not influenced by clocks. To make matters worse, the dominant trope of the study of time in anthropology was a sort of "us-and-them" approach that exoticized the other and implicitly left Western ideas as the standard against which all variation was described. My frustration led me to look at the literature on timekeeping in Europe that focused on the period before the dominance of the clock.

As I worked through the pre-clock textual record, it became clear to me that there was an omission in the literature on timekeeping. Histories that focused on Europe often presented a narrative of horological progress. Social scientific studies on other parts of the world told a tale of difference from Europe and struggles with adaptation and resistance. It was as if all diversity existed outside of Europe. I wanted to change that and write something that disrupted the logics of uniformity and a single timescale that have come to dominate both histories of European timekeeping and current time metrology. I felt that my previous work in a postcolonial context would be helpful in crafting this new project to challenge European timekeeping from within.

Four different strains of thought shape my approach. First, I was trained in psychological and cognitive anthropology, and this guides my interest in how people use objects in their environment to think—that set of ideas was a powerful guide in my book *Objects of Time*. Second, the social scientific approach to time that emerged in the 1990s has been particularly influential to how I approach the topic of time. Carol Greenhouse explored time pluralism,[2] and Barbara Adam documented the dissonance between lived rhythms and clock culture.[3] Their emphasis on plurality is the conceptual foundation on which my project is built. Third, my interest in cognition, objects, and pluralism has a natural resonance to the attention given to design by Michelle Bastian and Larissa Pschetz.[4] Bastian's idea of liberating clocks from their current design is a particularly powerful idea for me. Finally, the work of Caribbean thinkers such as Édouard Glissant, Wilson Harris, and C. L. R. James encouraged my view of the trajectory of European timekeeping as not only arbitrary but contrary to the experience of living on a rotating geoid.[5]

Notes

1. Birth, *Objects of Time.*
2. Greenhouse, *A Moment's Notice.*
3. Adam, *Timewatch; Timescapes of Modernity*; "Clock Time"; and *Time.*
4. Bastian, "Liberating Clocks"; Pschetz, "Time Affordances and Physical Mobility"; Pschetz, "Time"; Pschetz and Bastian, "Temporal Design."
5. Glissant, *Caribbean Discourse* and *Poetics of Relation*; Harris, *Tradition, the Writer and Society*; James, *Wilson Harris.*

Acknowledgments

I thank the participants in the "Science of Time" event held in June 2016 for their responses and comments on my paper. With regard to that paper, I thank Paul Gabor and Ken Seidelmann for their comments. That presentation is what grew into the present work. I am grateful to Charlotte Vignon and Joseph Godla of the Frick Collection for access to that museum's clock collection and for photographs of the Weber masterpiece clock. Demetrios Matsakis and Dennis McCarthy gave me valuable comments and corrections on the chapter about the leap-second controversy. The chapter on York began as a presentation at the "Time, Regulation, and Technoscience" conference in York, UK, the outlines of which I also discussed with Elizabeth Hartley and Peter Addyman. Owen Gingerich and Sara Schechner provided valuable leads for information about Regiomontanus and Nuremberg timekeeping.

The Jost Bürgi Library of the New York Horological Society has been of immense help, and I give a special thanks to Miranda Marraccini and St. John Karp for addressing some of my requests for obscure sources.

I thank Anthony Mason, Divjot Kaur, Caroline Kuhtz, and the rest of the production staff of Berghahn books for their making the publication process smooth, and Nicki Averill for addressing my quirky requests in crafting the wonderful cover art.

Finally, I thank my wife, Margaret, whose editorial skill in multiple languages has been extremely valuable in improving my prose and in correcting my translations.

Introduction

What Is Time Pluralism?

> Every age, including the postcolony, is in reality a combination of several temporalities.[1]
>
> —Achille Mbembe, *On the Postcolony*

On a late July day in 1858, Mr. March's court case was scheduled take place before judge Baron Watson in Dorchester, England. That morning, Judge Watson arrived at the court "punctually at 10 by the clock in the Judges' room, which was set at Greenwich time."[2] When he called the court into session, neither the defendant nor the defendant's counsel was present, so Judge Watson decided against poor Mr. March. Soon after, Mr. Cole, who represented March, entered the courtroom and protested Judge Watson's decision. Mr. Cole argued that the court had sat and made the decision before ten o'clock, the scheduled time for the trial. He then demonstrated that he had entered the courtroom at 10:00 AM according to Dorchester's town clock. This led to a discussion of what the true time was—Dorchester time or Greenwich time.

Dorchester is 2.4 degrees of longitude west of Greenwich. Before there were time zones, every degree of longitude represented a four-minute time difference. This meant that before Greenwich Mean Time became the time standard for Britain (which happened in 1880), the local time in Dorchester was about ten minutes behind the time in Greenwich—that is why the town clock and the clock in the judge's chamber disagreed.

To his credit, Judge Watson recognized the problem posed by Mr. Cole, but the plaintiff's attorney had already left the courthouse, so the case could not be retried at that moment. As a compromise, Judge Watson agreed to Mr. Cole's request to have the case heard by a higher court.[3]

In November, the case came before the Dorchester assizes. The court noted that such differences of public clocks indicating different times created confusion that had legal ramifications. Judge C. B. Pollock determined that the case should be retried without costs to Mr. March. He concluded his ruling stating, "Ten o'clock is 10 o'clock according to the time of the place." He further ruled the no town council or railroad could change the time of the place.[4] In other words, the legal time was the local mean solar time, not the mean time at Greenwich.

"What time is it?" This is a mundane question. Unlike that 1858 summer in Dorchester, today, few people dispute the answer to that question. This is because we live in a world where many carry electronic devices that synchronize automatically. In fact, this is very strange. At no other point in human history have clocks been so closely synchronized and consequently placed above question. Yet, disagreement between timepieces used to be common. For instance, the eighteenth-century poet Alexander Pope wrote:

> 'Tis with our *Judgments* as our *Watches,* none
> Go just *alike,* yet each believes his own.[5]

Now it seems that the time is one of the few things about which everyone agrees. Many clocks and watches set themselves even when there is a shift between standard time and daylight saving time, or when one travels across time zones. Technology makes our clocks and watches go alike. Clock users do not need awareness of how the time on the clock is determined. The time is beyond challenge, in part because everybody seems to have the same time represented on their clocks, watches, and devices.

Even as such synchronization requires little effort by clock users, it involves great effort among those who manage time metrology and time-synchronization practices. These individuals manage suites of atomic clocks, implement time-synchronization protocols, and shape policies for time standards. They seek to improve the precision and accuracy of the clocks they manage, as well as the precision and accuracy of the time available to all of us who rely on those clocks. When those who oversee current time standards, time metrologists, refer to precision, they refer to the measurement of small increments of time. When they refer to accuracy, they mean the ability of the clock to reliably keep the correct time. One can have an atomic clock that measures billionths of a second (nanoseconds), but if it is set incorrectly, it is still not accurate. Ideally, clocks should be both accurate and precise.

As scientists push for new levels of precision, this precision creates challenges, including a new permutation of Pope's point—clocks still do not run alike. This is not apparent when one looks at a clock, yet if one peels away the veneer of agreement, as this book seeks to do, one can see that clock time is not settled. Whereas in Pope's day, the differences might be minutes or even hours, now the differences are at the level of nanoseconds or less. While such tiny differences exist outside of the notice of most clock users, they are significant to those in charge of maintaining time standards, such as the Time Service of the International Bureau of Weights and Measures (BIPM).

The Time Service of the BIPM receives reports from forty-five national time laboratories across the globe. It uses these reports to calculate International Atomic Time (TAI),[6] which is then used to determine the globally distributed time standard, Coordinated Universal Time (UTC). The Time Service periodically publishes an online document called *Circular T*, which indicates how far off each time laboratory's clock is from UTC. Not only does *Circular T* indicate deviation from the standard, but it displays differences between these laboratories. On 26 September 2024, the clocks of the United States Naval Observatory were -1.6 nanoseconds behind the standard. The best-performing time laboratories on that date were the United States' National Institute of Standards and Technology, and the National Institute of Metrology in China, although a week before, the time laboratory of the Istituto Nazionale di Ricerca Metrologica in Italy performed the best. Such variation from week to week is one reason why no single metrological laboratory is relied upon by the BIPM.

Clock drift and uncertainty—these issues vex time metrologists seeking synchronization, but not many other people. For most people, all clocks are the same and represent an unquestioned clock time. Instead of making choices about time, many people accept the time on their clocks as ontologically given. This reflects an unquestioning faith in an assumption of an ideal clock that is always correct and keeps uniform, continuous time. The imagined ideal clock is so pervasive that even histories of timekeeping almost treat this idea as an inevitable outcome. In contrast to faith in the ideal clock, because of historically documented and contemporary multiple temporalities found in everyday practice, Larissa Pschetz suggests that there is a need to think about time and timekeeping in new ways.[7] She suggests that we have inherited a set of ideas about time ossified in objects, and that the ideas associated with the perfect clock are no longer designed to meet present needs. Instead, clocks become a technology that dictates behavior to suit their design. They suppress diverse social rhythms,[8] and I add that they obscure the diversity of environmental cycles.[9] The implication of Pschetz's point is that clock users adapt themselves to clocks rather than clocks somehow adapting themselves to users' needs.

By projecting the ideas of uniformity and standardization of the ideal clock onto material objects, material clocks gain remarkable authority. If one moves the other direction by relating material clocks needing synchronization to how the ideal clock is imagined, the idea of an ideal clock becomes unsettled. The use of the term "unsettled" is meant to emphasize the quality of a lack of stability and fixedness.

William James observed, "*Any object which remains uncontradicted is ipso facto believed and posited as absolute reality.*"[10] While the concept of the ideal clock fosters faith in clocks, in fact clocks are objects that materialize conflicts. They need to be reset, synchronized, or steered. Otherwise, they contradict one another. Awareness of that needs to be hidden or at least suspended in order for clocks to be believed as faithful representations of time. In fact, conflating the time of material clocks with an imagined ideal clock elides the messiness of timekeeping and grants an unwarranted authority to the material object. Henri Bergson did this in his debate with Albert Einstein over the issue of simultaneity.[11] For Einstein, a clock was enmeshed in its frame of reference—there were only clocks, and nothing like a perfect clock. The phenomenon of time dilation was a result of different relative velocities of different frames of reference.[12] For Bergson, every clock in every frame of reference theoretically ran at the same tempo for the clock user, and because of that, he denied Einstein's view of time as relative.[13] Deleuze sorted this out as a contrast of multiplicities—the multiplicity of measurements (Einstein) versus the multiplicities of the experience of subjective durations (Bergson).[14] The compromise between the two positions allows for clocks running at the same tempo, but for time measurement between frames of reference to be relative.

In this compromise, clocks' homogeneous, uniform time is a myth. It is not a myth in the sense of being false, but in the sense of a foundational story around which experience gets organized. It is a sacred story. Granting this, Bergson imagined a material homogeneity of performance of clocks that time metrologists do not accept.

In this context, how should the time standard be set? The time produced by these time laboratories varies in accuracy from one week to the next, so privileging one above the others makes no sense. On the other hand, if one uses an average, what should the algorithm be? Should it weigh all clocks equally, including poor-performing clocks? After all, on 26 September 2024, the Time and Frequency Lab at Cagliari was -16170.6 nanoseconds off UTC, and such variances could skew the average. Actually, in the face of clocks not running quite alike, the BIPM uses a weighted average, but even then, the determination of the relative weights of the different reporting time laboratories is political—whose time gets weighted the most? Do countries with multiple time laboratories, such as the United States and China, get to shape global time more than other nations? In effect, behind the apparent uniformity of clock time is a multitude of competing clocks indicating slightly different times that result in different opinions about how time standards should be defined and publicly distributed.

This is not just about measuring time. People use concepts of time to organize and regulate social activities. Power over time is power over people. Throughout most of human history, time has been debated and multiple ways of reckoning time have coexisted. Even with the advent of clocks and a seemingly agreed-upon timescale, disagreement between clocks has been used strategically to gain an edge in relationships.

For instance, a friend of mine in Trinidad relayed an anecdote that continues to provoke my thinking. He worked as a foreman supervising crews that maintained roads. One of his responsibilities was to keep track of the time to ensure that the workers under his supervision worked a full day. Each morning, he would set his pocket watch to that of the supervising engineer before taking his crew to the assigned location. He recounted that one day he decided to let everyone go home early, but shortly after, the supervising engineer arrived. The engineer accused my friend of letting the workers go home before quitting time, but when my friend pulled out his pocket watch and showed it to his supervisor, the watch indicated that the time was slightly after quitting time. As he did so, he added, "This is the time I work by." His boss could do little more than suggest that the watch get checked out.

The watch was a reliable timekeeper, but my friend had mastered the art of pulling out the watch's stem and changing the time displayed while the watch was still in his pocket. By using this trick my friend got the engineer to treat the watch as an authoritative source of time even when it disagreed with the engineer's own watch. As a result, my friend got away with letting his crew go home early while only receiving a rebuke from the engineer to get the watch checked. This is an example of time pluralism, a term coined by the anthropologist Carol Greenhouse to mean "the coexistence of multiple time constructions within social fields."[15]

My Trinidadian friend's story stands in contrast to the rhetoric, but not the reality, of current time standards. These standards and the infrastructure used to maintain them seek to eliminate the possibility of multiple, coexisting clock times—a form of temporal pluralism tied to time-keeping technology. Instead, the goal is a single, uniform, universal time that clocks reliably display. Cross-culturally and historically, time pluralism has been the most common state of affairs. The uniform, standardized system under which the world currently lives is a recent innovation, if not an aberration.

Today, it is hard for many people to imagine a society without synchronized time indicators, much less a society that employs multiple, different ways of reckoning time. On the other hand, as Emily Keightley demonstrates, when one focuses on the human agent rather than the technology, the time provided by technology does not suit many experiences of time enmeshed in local power relations.[16]

A common example is jet lag. One can fly across many time zones and one's smartphone will synchronize with the time zone where one lands, but one's body does not. Humans are diurnal mammals with pronounced circadian cycles; clocks are examples of a technology that strives to be continuous and uniform.

The homogenizing of time across modern nation-states hides the reality that most of the world's population lives under some form of time pluralism. The most obvious examples are calendars. The Jewish, Muslim, Hindu, and Chinese calendars differ from the "secular" Gregorian calendar (a bit of an oxymoron when one considers that the calendar was imposed by the Roman Catholic Church and only later adopted by the world). Some societies are bi-calendrical. The anthropologist Janet Hoskins observes that Kodi ritual calendars used on the island of Sumba in Indonesia are at odds with the calendar of the Indonesian state.[17] The Ottomans kept both an administrative solar calendar and the Islamic lunar calendar.[18] In the use of the Gregorian calendar, Jews, Muslims, Hindus, and Chinese see their holidays as "moving" from one year to the next. Truth be told, their holidays do not move. For instance, every year Passover is the fourteenth day of the month of Nisan. It never changes. If history had been different, and the Jewish, Muslim, Chinese, or Hindu calendar had been dominant, New Year's Day in the Gregorian calendar would move in the dominant calendar. The majority of the world's population lives with time pluralism and refers to multiple calendars, and when combined, there are more Jews, Hindus, Chinese, and Muslims than there are people who use only a single calendar.

This book focuses on the topic of time pluralism in daily timekeeping, and does so by looking at examples from the past and the continued existence of such pluralism. Current clocks are deceptive in their apparent uniformity and synchronization, but such uniformity and synchronization is fragile because of several unresolved questions. Can a single time standard be imposed on the world? Should a single time standard be imposed? Is time pluralism really a threat? Looking at time in this way undoes the standard tale of progress from early clocks through the atomic clock but instead looks at the reality that time measurement is a consequence of human choices about techniques, methods, and technologies. At different points in the past, these choices have allowed for multiple public times. At other points, such as now, diversity has been suppressed. There is benefit in revisiting these past examples of time pluralism to think about the present time system and its future.

Every Clock Tells a Story

Every clock tells a story that concerns a debate about time; every clock materially represents ideas; thus, every clock attempts to shape how people think about temporality in a context of debates about time. The stories here are about the articulation of clockwork and clock displays. A clock is a device that converts transitions (ideally of equal duration) into culturally recognized units of time that are then displayed. The following chapters will show that it is possible for very different cultural units of time to be represented using the same clock mechanisms. Clocks differ from clepsydrae and sundials, which work based on continuous processes, like changing water levels or the movement of a shadow. Such continuous movement then crosses marked thresholds, such as hour lines. It is the crossing of those thresholds that marks time. Consequently, there are two very different logics embedded in these devices. Clocks work on the principle of an addition of discrete units that are converted into culturally recognized categories. Clepsydrae and sundials work on the principle of a continuous process crossing thresholds labelled as culturally recognized categories.

The sequence of culturally recognized units is a timescale. What links a clock to the idea of uniformity and continuity is the concept of a timescale. A timescale consists of the assumptions, logic, and units for measuring time. Timescales determine representations of time, and how those representations are achieved. A clock is not a timescale, and a timescale is not a clock. A timescale is a set of ideas to which a clock's performance is compared. UTC is the current globally distributed standard timescale, and there are no clocks that authoritatively display UTC, only local close approximations of UTC. That said, an ideal clock would display UTC.

Louis Essen, one of the inventors of the atomic clock, wrote that timescales serve three functions: "of giving the time of day, the season of the year, and also a measure of time interval or duration."[19] For a timescale to serve these purposes, it must have the following features: 1) a starting point that defines the epoch to which the timescale applies, 2) units of time, and 3) a sequence of the units that extends from the beginning point into the future or to a point where the timescale repeats. These ideas are essential for the definition of the time of ideal clocks.

It is not clear when the term *timescale* first emerged, although its importance as a concept seems to coincide with the diminishing public knowledge of how clock time was determined. The *Oxford English Dictionary* gives the first usage as occurring in 1828. Another indication that it is a relatively recent term in English is the fact that there is not yet a standard spelling—it can be spelled as *time scale* or *timescale*. While the term *timescale* seems to be relatively recent in origin, it is remarkable that it is used so frequently without definition. It is as if the term is utterly noncontroversial.

On the one hand, the 1828 date the *Oxford English Dictionary* gives is plausible because the idea of a timescale seems closely tied to ideas of standardization of time across space and the emerging idea of a reliable, uniform, continuous clock associated with modernity. Such ideas emerged among metrologists in the eighteenth and nineteenth centuries. Yet, timescales preexist 1828. The most obvious examples of timescales come from chronologies where the creation of a single scale of time, in order to represent the sequence of events, is important. Notable examples of this include the Islamic Hijri chronology or Bede's use of Anno Domini. Bede's case is particularly instructive. He starts out his book *The Ecclesiastical History of England* using timescales associated with the reigns of particular monarchs,[20] but since Anglo-Saxon England involved several kingdoms, and Bede's historical vision included the entire Church, such timescales based on single rulers were unwieldy. This is possibly why by chapter three of his book he begins to use the idea of Anno Domini in order to chart events across kingdoms and space. It seems that once one thinks beyond the local, some form of standardization across space becomes desirable.

The existence of the term timescale in English itself raises questions. It is a European term, most likely of either French or English origin. Moreover, in debates about timescales, the languages of choice have been French or English. This is not because the concept to which the term timescale refers does not exist in other languages; it is just that those other languages use words that are compounds of "time" and "scale," as well, such as in German where it is *Zeitskala*. Some languages, like Japanese, borrow the term from English, with the Japanese word for *timescale* written in either katakana as *timu sukeru*, or written with a combination of katakana and kanji: 時間スケール (*Jikan sukeru*—literally "interval of time scale" with *sukeru* being the Japanese spelling of the English word "scale").

Timescale representations are cultural creations out of social processes, which means clocks are results of those processes. One such story can be read on any current clock or watch—the story of the multicultural origins of our current way of displaying hours. We inherited dividing the day into twenty-four hours from the ancient Egyptians. The sexagesimal system for minutes and seconds originated in the Near East and diffused with the help of Alexander the Great. The division of seconds into tenths, hundredths, thousandths, and so on, is a legacy of the decimal time system of the French Revolution. The representation of time seen on clocks is not an invention of the West, then, but from its Renaissance origins, it was an early example of what the philosopher of science and technology, Don Ihde, called "postmodern pluriculture"[21]—a combination of borrowings from several cultural traditions. The combination is Western; its components are not. This cobbled-together collection of ideas about time makes up the timescale represented on clocks. In a way, it is a reconciliation of different ways of counting time units—a resolution posed by previous problems of time pluralism.

That simple display mediates a relationship between the clock user and millennia of people thinking about and debating time. It implies a story of how these different time systems came together and the choices that brought them together. Clocks mediate between expert knowledge and users,[22] and any clock one looks at includes multiple generations of experts.

These experts are hidden from view. Most people cannot name the person in charge of the Time Service of the International Bureau of Weights and Measures (BIPM), or the heads of time laboratories spread across the globe. While the day is a fundamental unit of time for many people, few know that its current recognized length was determined by Simon Newcomb.[23] Indeed, more people know the fictional character based on Newcomb, Sherlock Holmes's arch-nemesis Professor Moriarty, than Newcomb himself.[24]

The lack of knowledge is because most users of clocks end their education about timekeeping early in primary school. The educational goal is not a comprehensive knowledge of timekeeping, but merely how to "read" a clock. This education muddles an imagined ideal clock that keeps uniform and continuous time with what material clocks actually display. The relationship between the imagined ideal clock and actual clocks is taken for granted. From this perspective, such basic education in timekeeping promotes an unquestioned association of the ideas of time associated with the ideal clock and how actual clocks represent (or misrepresent) that ideal. Many people treat clocks that synchronize automatically as faithful simulacra of the ideal clock.

The globalization of time standardization embedded in information and communications technology has a logic different from material clocks—those individual devices that people use to tell the time.[25] The ideal clock keeps time perfectly; clocks get set. The ideal clock is uniform in its timekeeping; clocks break and malfunction. Bruno Latour observed in *Aramis,* his study of the demise of plans for a rapid transit system in Paris, that the malfunctioning of technology influences the politics of technoscientific decisions.[26] It is often these malfunctions that drive changes in the efforts to get time-keeping technology close in function and representation to the ideal clock.

The material clock mediates cognitive relationships between users who do not question working clocks, designers/builders of clocks, and time policy makers who worry about the variability of clocks.[27] When one shifts perspective from the homogeneity of the ideal clock to the diversity of actual clocks the picture becomes far more complicated. As Paul Glennie and Nigel Thrift point out, "clock time comprises a number of concepts, devices, and practices which have meant different things at different times and places, and even in any one place have not had a unitary meaning."[28] Such differences are what almost got Mr. March convicted in 1858 and allowed my friend to go home early without a reprimand from his supervisor. Instead of viewing clocks as reflecting the ideal clock in a simple, straightforward manner, we should view clocks as material objects positioned in a contentious and complicated chain of ideas about time and its measurement. The current relationship between the ideal clock and clocks works against public knowledge about the variation of clocks.

When looked at in terms of scientists thinking about time and encoding their ideas into clocks, it is not surprising that different clocks can indicate many different ways of thinking about time. Actual clocks are devices designed *and imagined* to pursue the chimera of abstract, ideal timekeeping. The ideal clock is uniform and consistent; clocks are a means, often flawed, of conveying that idea to users. Despite being an abstraction, the ideal clock depends on the material limitations of technology. But in this relationship of the abstract and material, most clock users overlook the flaws and the diversity of the materiality of clocks even as such flaws and diversity are a source of concern to those who oversee the definition and distribution of UTC.

Humans' use of clocks diverges from the time on the ideal clock, however. The time of the ideal clock is uniform and continuous; the human consultation of clocks is momentary and discontinuous. Added to this constellation is the use of clocks in information technology, such as computers timestamping changes to files. This is a documentary function of clocks that is expected to run outside of the intervention of users but to be available to users for consultation. To know the exact time at any given moment depends on a device that measures out increments in a continuous timescale—something that humans are incapable of doing without the aid of technology.

Time Prosthetics

People, including time metrologists, are incapable of knowing the correct time in the current time system without the aid of a clock. Because of this, the anthropologist Laura Bear describes the clock as a prosthetic for knowing the time.[29] Such prosthetics are part of what makes us human—we rely on objects to manage our world and our experiences. Objects that define our humanness not only reflect culture but are essential components of many cultural processes.[30] Louis Leakey wanted to make tool making and tool use a defining feature of the genus *Homo*.[31] From Leakey's perspective, this norm goes back millions of years and defines our humanity. Yet, as Bill Brown argued in his article "Thing Theory," there has been a tendency to separate objects from ideas,[32] and even to neglect objects in favor of the mental, rather than to confront the complicated relationship between objects and ideas. Indeed, one of my mentors derided the term *material culture* with the glib comment, "Material does not have culture."

Since the 1980s, there has been resistance at such separation of things from thinking and social relations.[33] Within cognitive science, in part because of the idea of cognitive tools in L. S. Vygotsky's work,[34] the idea of using elements of the world or material creations to serve as extensions and augmenters of cognitive processes has gained attention.[35] We use clocks to know the time; this makes them what Edwin Hutchins calls *cognitive artifacts*.

Cognitive artifacts are tools for thinking[36]—they are things that perform cognitive tasks for us. Humans augment their mental capacities by creating tools that complete cognitive tasks. For instance, a calculator is a means of doing arithmetic that one might be able to do in one's head or with pencil and paper, but it's faster and easier to use the calculator. A cognitive artifact can be as simple as taking notes for future use rather than relying on memory alone, or as complex as a computer doing complicated analyses of large data sets.

Because of the knowledge encoded in a cognitive artifact, the use of these objects involves little or no reflection on how their results are produced.[37] Much of modernity relies on schedules, and to know the correct time associated with those schedules, we rely on machines—our clocks and watches, the devices involved in transmitting time synchronization protocols, and ultimately atomic clocks. Nobody, not even time metrologists, can know the time by the direct observation of the world around them. So our timekeeping is a fusion of the human judgement of time metrologists and their atomic clocks, systems administrators and their computers and clocks involved in synchronization protocols, and clock users. Clocks are artifacts that link these people and processes. When it comes to knowing the time, we are cyber-linked to a host of time-keeping experts and technicians through the cognitive mediation of an object. It is impossible to know clock time without this complex network of material clocks and the people who manage them.

Such dependence leads to forgetting the choices and ideas that went into the design of the objects.[38] The clock becomes not just the source of expert knowledge, but the expert itself. Clocks hide the social relationships between clock users, those who maintain time standards, the designers of the current standards, and, ultimately, the relationship to the ideal clock.

André Leroi-Gourhan saw this exteriorization of thought into objects as a new stage in human evolution,[39] adding: "The first means of programming by purely mechanical processes were found in clock-making."[40] Clocks are an example of such exteriorization, and they are a significant example because they pervade so much of everyday life.

To some extent, for the prosthesis to work, there must be blind faith in clocks' representation of time. A psychology experiment in the 1960s demonstrated the credence given to clocks. The subjects were asked to do a simple reading task and record their time by a clock on the wall, but they did not realize that the experimenters manipulated the clock's rate to run slower than normal. When the clock was running at 50 percent of its normal speed, the subjects still regarded it as running normally. This was also true at 25 percent of normal speed. It was only when the clock was running at 12.5 percent of its normal speed that the subjects began to suspect that the clock was being manipulated.[41] Such unwavering faith in clock performance even when clocks perform poorly allows them to regulate the lives of the faithful. Clocks regulate and structure social relations, labor, and even body functions, thereby crafting what Elizabeth Freeman calls "chrononormativity."[42]

It is an idea of great power tied to ubiquitous devices. As an accessory to faith in the ideal clock is the argument that Larissa Pschetz and Michelle Bastian put forth: that material clocks shape the experience and perception of time.[43] Additional work on speeding up or slowing down clocks reveals that its effect on time perception has additional effects over the experience of fatigue, hunger, and pain.[44] Clocks are prosthetics by means of which we know the time, and clocks determine and define the time we know, occluding other possibilities.

When the idea of the clock as prosthetic is applied to the idea of time pluralism, one recognizes the threat to standardization. It is not just the clocks that differ, but the people who use them. Going back to the story told by my Trinidadian friend, his supervisor had to believe my friend because they both chose to believe my friend's watch. Without the watch (and the trick of resetting it), my friend would not have been able to justify letting his crew go home early. Because clocks are necessary prothestics for reckoning time, the trend toward clock standardization is also a trend toward the standardization of the human experience of time.

Necromantic Prostheses

The idea that clocks standardize people has a sinister ring to it, but it does not indicate some secretive cabal of mad scientists seeking to control our thoughts. Clocks embody ideologies. Since they are things that people rely upon to think, these things "act." Nomi Lazar demonstrates that they are also devices that are used to establish and maintain political legitimacy.[45] The common description of clocks is that "clocks tell time for people"—in the syntax of that phrase, the clocks are the agents and people are objects of agency.

Going back to the idea that every clock tells a story, most clocks tell stories with long histories and multiple generations of both named and nameless scientists, designers, and policy makers. To put it bluntly, much of the knowledge mediated by clocks was generated by people long dead. As I argued in *Objects of Time*, this makes clocks necromantic objects—a means by which the knowledge developed by dead people is communicated to the living.[46] In this situation, one can no longer assume that clock time is what it is because of its function or some conscious planning of a small group of individuals in the present. Each generation of time metrologists inherits the technology of previous generations, and in so doing, inherits the ideas embedded in that technology. Yet, since clocks reproduce concerns of the past, the politics embedded in clocks are not necessarily those of the present. Time standardization creates products of this inheritance.

For this reason, outside of the conscious manipulation of clocks, the technology dictates behavior, but behavior does not dictate the technology—human beings are "locked into the material and technological world they create."[47] Clocks are the way they are because of an alchemy of the reproduction of past designs through habit and present uses. That combination of ancient Egyptian, Babylonian, and French Revolutionary ideas did not occur because those ideas were good to combine. Instead, the combination occurred because it represents a history of accumulated habits.

Cultural knowledge is distributed across individual humans, but is also distributed through objects that transcend generations and cultures, and with current telecommunications technology, transcend space, as well. In fact, often the designers of time-keeping knowledge and technology die long before their technology ceases to be used. The list of such people includes the famous, such as Galileo; the less famous, such as Louis Essen, one of the inventors of the atomic clock; and the unknown, such as those who toiled on factory floors faithfully reproducing clock designs. Davis Baird points out about scientific instruments in general that this leaves their inventions in the hands of technicians: "Expert knowledge systems transcend their makers."[48]

The ideas of dead scientists and metrologists embedded in clocks shape our time consciousness.

Because of reliance on these necromantic cognitive artifacts, habit becomes more powerful than science: Lazar notes that "in the absence of renewed conflict, temporal order grows familiar and may pass beyond notice."[49] Such a melding of habit with past and present maintains current information technology infrastructure. Clock users act using cognitive artifacts that function based on preprogrammed algorithms. An example is the time synchronization protocols of computer systems. These protocols are not clocks but facilitate the agreement between clocks. Once set up, something like Network Time Protocol (NTP)[50] is designed to robustly synchronize computers through communicating with time servers with little or no human intervention. This allows NTP to frequently and reliably synchronize computer clocks, which traditionally have kept time poorly, to give the appearance of computers displaying an accurate time.

To give an example, several years ago when I proctored an examination, I projected onto the classroom's screen two clocks: the "official" time of the United States provided by the National Institute of Standards and Technology (NIST), and a dial showing a second hand representing the desktop computer's clock. At the beginning of the test, there was a 17-second difference between the two clocks. At some point during the examination, the computer's clock synchronized with a time server, and it started to show no difference from the NIST's time. Such events occur many times a day outside of the awareness of most users of electronics.

Clockmakers and time metrologists did not create NTP. Computer scientists who designed it desired a protocol for using networks to distribute time signals generated by those who manage high quality atomic clocks. So, in addition to clock designers and makers, in this web of transgenerational and translocal relationships, are also those engineers and technicians who maintain clocks as mediators. In effect, the production and exchange of clock time follows the logic of what Ian Hacking observed about laboratory sciences: There is an interconnection of the things themselves, the facilitators and makers who instruct users on how the things are to be used, and the technicians who work with users to keep the things in working order.[51]

The consequence of this necromancy is that the concerns of the designers and facilitators, some of whom are long dead, sometimes differ from those of the users, and as a result, inherited designs embody cultural logics of bygone eras. The modern clock is a good example. The clockwise movement of the hands mimics the movement of the Sun's shadow on a sundial in the northern hemisphere—the hemisphere in which clocks originated. In charting the movement of this shadow, analog clocks adopt a geocentric representation of the relationship between the Earth and Sun in the northern hemisphere. We know that the "movement" of the Sun across the sky is a product of the rotation of the Earth, and not of the Sun's motion. Moreover, the shadow on a sundial moves in a counterclockwise direction in the southern hemisphere. An accurate clock representation of Earth's rotation that did not privilege one hemisphere over the other would have the hands stay stationary and the dial move in a counterclockwise direction. This sort of display has been tried, such as in the decimal clock made by Abraham Breguet that is now at the Frick Collection,[52] but such representations never caught on. The northern hemispheric geocentric display is far too ingrained in clock users' habits to change. In effect, the cognitive artifact leads the users to imagine cycles incorrectly.

Clocks and calendars are essential equipment for the durability of many institutions.[53] The persistence of knowledge after death that gets embedded in the design of common cognitive tools, like clocks, has a conservative quality to it. Normally, technology is viewed as an agent of change, but objects can also be agents of stasis, particularly when humans' habits are tightly integrated with the object. This gives rise to another source of time pluralism—the differences created between the primary school knowledge of time deeply embedded in habit and the advancements in time-keeping and time-synchronization technology and practices.

Michelle Bastian argues that this undermines the tendency of philosophical discussions of time to see clocks as "self-evident tools" that "are not thought to be worth remarking on."[54]

Basically, clocks are in a largely unexplored black box[55] with a largely unexplored relationship to the idea of the ideal clock with its ideal uniformity and homogeneity of time units. Michel Callon and Bruno Latour describe black boxes as containing "that which no longer needs to be considered," and the more that can be placed into a black box, including habits and modes of thought, "the broader the construction one can raise."[56] For a globalized network of information and economic flows that the ideal clock provides through clocks, and all the transgenerational and translocal knowledge clocks encode, the black box needs to be impressive, indeed. Despite all the writing about clock time, in most cases clock time is still, as Michelle Bastian suggests, "an unassailed monolith."[57] What is inside that box?

The history of timekeeping has been one of different ways of counting hours and different clock designs; current clocks obscure time pluralism through such necromancy. The efforts to create and maintain a standard, uniform timescale is a fight against generations of habit and time-keeping technology. This turbulent history must be suppressed and pushed out of consciousness for official clock times to maintain their legitimacy. It is this history and these politics that are in the black box.

The Diversity of Clocks

I argue that the ideal clock is an accumulation of ideas and stories reproduced by habit and incorporated into experience. Its representation of time could have been otherwise. As the following chapters discuss, contrary to the impression that there is only a single time clocks can represent, the history of these devices reveals that the development of mechanized timekeeping produced a proliferation of different ways to represent hours.[58] The neglect of these other systems of representing time in histories of timekeeping cultivates the impression that there is only one "natural" clock time.

Peter Burke points out that "the rise of the mechanical clock . . . no longer looks as simple a phenomenon as it once did."[59] The simplistic view of the history of clock time emphasizes the standardization of clock hours that has taken place since the Enlightenment. If one reads histories of clock time written before 1900, one does not find a simple story of a single form of clock time emerging out of some inevitable march of horological progress. On the contrary, one encounters different hour systems and differences between mean time (a time that divided the average length of day into twenty-four hours) and apparent solar time (a time based on the angle of the position of the Sun to one's meridian on a specific day).[60] Because of the present delusion of simplicity in the development of the clock, there is little recognition of our living at a unusual moment in concepts of time—a moment of unprecedented standardization with a veneer of global agreement.

Such histories also tend to neglect the relationship of colonialism and time standardization.[61] Global imperialism used standardization to suppress multiple ways of conceptualizing time. This did not eliminate alternatives but established the European ways of reckoning as dominant.[62] One should not assume that the movement of Western material culture indicates assimilation to Western ideas, however.[63] Western clocks were vectors for communicating Western ideas of timekeeping, but clocks could be adapted to local, non-Western ideas. There is a growing body of literature on how other ways of reckoning time have survived the domination of the "Western" ideal clock. Such cases have been documented for Egypt,[64] Turkey,[65] and Japan,[66] and the literature is bound to expand as this issue becomes more widely explored.

Consequently, I disagree with the position that "other times—non-Western and pre-modern—become silenced through a form of temporal imperialism."[67] Instead, as the well-documented examples from Egypt, Turkey, and Japan demonstrate, globalization only increased competition between time systems.[68] Gerhard Dohrn-van Rossum argues that the current intertwining of horological technology and social habits should not be viewed as an outcome of Euro-American superiority, but of Euro-American arrogance,[69] and I would add that such arrogance gets in the way of seeing the diversity of time systems that continue to be used.

Instead of accepting and exploring a cultural diversity in clocks, the tendency has been to contrast "clock time" and "natural time." This binary distinction is pervasive and used to divide the world into "traditional societies" with imprecise and local times and "modern societies" with uniform and abstract time.[70] This is misleading in two ways. First, it essentializes all the cultural approaches to thinking about time that are not tied to the European science of time measurement, even if the non-European approaches include time-measurement devices. Second, it misconceptualizes modern timekeeping as beyond challenge, when, in fact, it is messy, contested, and plural, as chapters 6 through 8 will show. In this approach that focuses on the logic of the ideal clock, clocks are not sufficiently examined, and the diversity of times behind clocks is not explored.

Standards, Heterogeneity, and Globalization

Time metrology was and is a critical part of the architecture of empire and capitalism—in Latour's words, it has expanded "*locally everywhere*."[71] In addition to using clocks to coordinate social relations, clock time is a crucial element of the representation of the movement of capital by means of accounting and bookkeeping. It coordinates electric power grids. It is a necessary component of all computing in which marking actions with timestamps is essential. For all these time-reliant technologies and techniques to work, time must be standardized. As such, the measurement of time from the European metrological perspective became separated from locality,[72] even as many people's daily lives time remained intrinsically tied to their location.[73]

Guided by the concept of the ideal clock, time metrology created measures of time accepted and understood by all across large distances,[74] and thereby it became a global means of regulation and governmentality.[75] Pierre Bourdieu points out, "One of the first acts of civil bureaucracies … was the establishment of public time."[76] Clock towers were "structures that helped express the power of the government at a distance; they became architectural proxies for the state itself."[77] Such public time inclines authorities toward standardization, which in the case of European colonialism involved relating technology, mathematics, science, and chronology.[78] Time pluralism is not just a threat to standards, but to power.

Time standards have become an important component of the transnational informational infrastructure of capitalism, and with this, they have moved beyond the ability of any one nation to control.[79] In his work on the development of standardization, Laurent Thévenot notes, "Standards equip entities with a rather public kind of guarantee with the properties of objects,"[80] and in quantifying and governing so much of life, standards are also political.[81] Despite the importance of metrology in modernity, it is an under-researched topic,[82] and one in which metrological practice is often too closely tied to state actors rather than described as defying such actors' attempts at control.[83]

The distribution of standardized timekeeping did not result in the homogenization of all timekeeping, however. For instance, in Islam local timing is critical to prayer times and fasting during Ramadan even if the devout use a clock to know when to pray and fast. In such cases, the delocalized global time standard became harnessed to local timing—an example of what Partha Chatterjee noted in stating, "What was necessary was to cultivate the material techniques of modern Western civilization while retaining and strengthening the distinctive spiritual essence of the national culture."[84] For Islamic prayer times, apps now use local solar time derived from ephemerides represented in terms of local clock time. Concepts of time also shaped how administrative functionaries structured and represented their activities.[85] The hegemony of time standardization and uniformity that colonialism cultivated was and is incomplete, leaving many unexamined temporal tensions and conflicts that continue to be obscured by the image of the ideal clock.[86]

These factors cultivated habits that persist after decolonization. Now empires have dissolved, but the legacy of the ideal clock is a remnant of those empires. Because of the tendency to merge how the ideal clock is imagined with material clocks, as a result of postcolonial politics, when time pluralism is discussed, it is posed as an alternative, or even as a threat, to standardized clock time.[87] The universalizing aspects of time standards associated with the ideal clock conflicts with nationalist sentiments.[88] The course of postcolonial time metrology as a hybrid of European and non-European ideas combined with administrative habits has not fully made itself felt other than in the inability of any former empire to impose its metrological will on the United Nations' Radiocommunication Sector of the International Telecommunication Union (ITU-R), the international treaty-based organization that asserts responsibility for defining global time standards. The relationship between the ideal clock and clocks in general is set to become strained, and the relationship of time and experience is potentially complicated (or maybe liberated) because of time pluralism. This leads to a choice: suppress the diversity or be open to new clock designs that represent time pluralism and allow conversion between different ways of imagining time. The argument here is that the representation of time pluralism occurred in the past and is an option for the future.

The Weave of the Book

Time metrology seeks to quantify time, and as an effort at quantification, its logic is to reduce plurality and justify policy.[89] I argue that time pluralism is a common element of timekeeping. Starting with chapter 1, the book addresses cases of time pluralism in the past when it was more publicly recognized than in the present. Chapters 2 through 5 present historical cases of time pluralism. Chapters 6 through 8 show that, despite the present uniform appearance of clock time, challenges of time pluralism persist. What these chapters make clear is that the current choices in time metrology see time pluralism as a threat.

A general theme throughout is to acknowledge the possibility of multiple forms of clocks and clock times. If there is no necessary relationship between the time on the clock and any phenomena in the world, there is no reason to assume that clocks must have only one form.

Since every clock tells a story, there are clocks that tell stories of conflict, time pluralism, and conflicts over how the clock time should be imagined and other stories of competing clocks. How can one recover those stories?

To examine time pluralism is to examine different potential times that the ideal clock does not represent. In this approach, the representation of time becomes a site of manipulation and contestation by a few culturally knowledgeable people with implications for many others who take the representation for granted. The historian of medieval science, Lynn White, observed that "technology and science are the essence of power."[90] In the case of representations of time, there is yet another element that magnifies the power at the same time it hides it, namely, the power to manipulate a representation that most people have naturalized. A focus on multiple social rhythms complements the examination of time pluralism. This can be found in the examples of market regulation in medieval York (chapter 2), the relationship of clocks and gender among early modern Habsburg elites (chapter 3), and the emergence of modern astronomy through the timekeeping practices of Regiomontanus, Bernard Walther, and Tycho Brahe (chapter 4).

Finally, there is a political element to the manipulation of the representation of time. This theme runs through all the historical examples and culminates in the cases of the New York State Democratic Convention of 1858, and the enforcement of liquor laws in England discussed in chapter 5. In both cases, there is an agreement on the technology for measuring time and the representation of time, but there is also political manipulation based on the standard to which the technology and representation are set. In effect, there is a debate over which timescale is the timescale of the ideal clock and the timescale to which material clocks should aspire.

These historical examples are to address a concern expressed in current time metrology and highlighted in the debate over the leap second, as discussed in chapters 7 and 8. In this context, the relationship between clocks and the imagined ideal clock is particularly tight, and timescale pluralism is seen as a threat to the stability of this relationship and consequently a threat to the global technological infrastructure building on this relationship. In effect, the motivation to explore historical examples of timescale pluralism in the European horological tradition is to think about the questions of "what if" in debates about current time standards and to loosen the relationship between material clocks and the ideal clock. This is meant to disrupt a narrative of inevitable horological progress that assumes that future clocks will simply be improved versions of present clocks.

Notes

1. Mbembe, *On the Postcolony*, 15.
2. Hurlstone and Norman, *Exchequer Reports*, 719.
3. This case was covered by several English newspapers, with the earliest report I found being 23 July 1858 in *The Globe* on page 4.
4. Hurlstone and Norman, *Exchequer Reports*, 719–20.
5. Pope, *Essay on Criticism*, 4, emphasis in original.
6. By convention, the abbreviations for global metrological standards are based on French spellings. For example, while TAI is referred to as "International Atomic Time," its abbreviation is based on "*temps atomique international*."
7. Pschetz, "Isn't It Time To Change."
8. Pschetz, "Isn't It Time To Change," 60.
9. Birth, *Time Blind*, chapter 2.
10. James, *Principles of Psychology*, 289, emphasis in original.
11. Canales, *Physicist and the Philosopher*.
12. Einstein, *Relativity*, chapter 9.
13. Bergson, *Duration and Simultaneity*.
14. Deleuze, *Bergsonism*, 80.
15. Greenhouse, *Moment's Notice*, 96.
16. Keightley, "From Immediacy to Intermediacy."
17. Hoskins, *Play of Time*.
18. Wishnitzer, *Reading Clocks*; "At Approximately Eleven."
19. Essen, "Time Scales," 161.
20. Bede, *Ecclesiastical History*.
21. Ihde, *Postphenomenology*, 28.
22. Bear, "Time as Technique," 492; Birth, *Objects of Time*, 19–20.
23. Newcomb, *Astronomical Papers*.
24. Schaefer, "Sherlock Holmes."
25. Hassan, *Chronoscopic Society*, 7–8; Hassan, *Information Society*, vii–x, 25; Hörning et al., "Do Technologies have Time?"

26. Latour, *Aramis*, 68–71.
27. Birth, *Objects of Time*, 170–71.
28. Glennie and Thrift, *Shaping the Day*, 9.
29. Bear, "Anthropological Futures," 143.
30. Miller, *Material Culture and Consumption.*
31. Leakey et al., "New Species."
32. Brown, "Thing Theory."
33. Latour, "Actor-Network Theory"; Latour, *Reassembling the Social*; Latour and Woolgar, *Laboratory Life*; Appadurai, "Introduction"; Stiegler, *Technics and Time, 1*; Stiegler, *Technics and Time, 2*; Stiegler, *Technics and Time, 3*; Leroi-Gourhan, *Gesture and Speech.*
34. Vygotsky, *Collected Works of L. S. Vygotsky, Volume 1*; *Collected Works of L. S. Vygotsky, Volume 3.*
35. Hutchins, *Cognition in the Wild*; Hutchins, "Cognitive Artifacts"; Hutchins, "Role of Cultural Practices"; Hutchins, "Cultural Ecosystem"; Clark, *Microcognition*; Clark, *Supersizing the Mind*; Clark and Chalmers, "Extended Mind."
36. Hutchins, "Cognitive Artifacts."
37. Baird, *Thing Knowledge*, 19.
38. Baird, *Thing Knowledge*, 14.
39. Leroi-Gourhan, *Gesture and Speech*, 252; see also Stiegler, *Technics and Time, 1.*
40. Leroi-Gourhan, *Gesture and Speech*, 249.
41. Rotter, "Clock-speed as an Independent Variable."
42. Freeman, *Time Binds*, 3.
43. Pschetz and Bastian, "Temporal Design."
44. Thönes et al., "Cognitions about Time."
45. Lazar, *Out of Joint*, 3.
46. Birth, *Objects of Time*, chapter 2.
47. Hörnung et al. "Do Technologies have Time?" 294.
48. Baird, *Thing Knowledge*, 14.
49. Lazar, *Out of Joint*, 130.
50. Mills, *Computer Network Time Synchronization.*
51. Hacking, "Self-Vindication," 42.
52. Birth, "Breguet's Decimal Clock."
53. Latour, "Technology as Society"; Latour, *Never Been Modern*, 120–22; Lazar, *Out of Joint.*
54. Bastian, "Liberating Clocks" (unpublished manuscript, n.d.), 3–4.
55. Porter and Stockdale, "Strategic Manipulation," 273.
56. Callon and Latour, "Unscrewing the Big Leviathan," 285.
57. Bastian, "Liberating Clocks" (unpublished manuscript, n.d.), 2.
58. See Gustav Bilfinger, *Mittelalterlichen Horen*, 142.
59. Burke, "Cultural History of Time."
60. See Berthoud, *Histoire de la mesure*, and Bilfinger, *Mittelalterlichen Horen.*
61. Nanni, *The Colonisation of Time.*
62. Hassan, "Timescapes of the Network Society"; Postill, "Clock and Calendar Time."
63. Thomas, *Entangled Objects*, 186.
64. Barak, *On Time.*
65. Wishnitzer, *Reading Clocks*; Gürbüz, "Turkish Time of Yesteryear"; Gürbüz, "Ezani Time."

66. Conrad, "Nothing is the Way"; Fernandez and Fernandez, "Precision Timekeepers"; Foulkes, *Time Tamed*, 120–27; Frumer, *Making Time*; Mody, *Japanese Clocks*; Robertson, *Evolution of Clockwork*; Yokota, "Japanese Clocks and Karakuri."
67. Fitzpatrick, "Social Policy and Time," 202.
68. Mawani, "Law as Temporality."
69. Dohrn-van Rossum, "Uhren und Uhrzeiten," 77.
70. Burke, "Cultural History of Time," 619.
71. Latour, *Reassembling the Social*, 229, emphasis in original.
72. West-Pavlov, *Temporalities*, 123.
73. Ogle, *Global Transformation of Time.*
74. Uribe, *Las Dinámicas del Tiempo,* 19.
75. McCrossen, "Conventions of Simultaneity."
76. Bourdieu, *On the State*, 8.
77. Rooney, *About Time*, 23.
78. Adib and Emiljanowicz, "Colonial Time in Tension," 1,225.
79. Hope, "Conflicting Temporalities"; Hope, *Global Capitalism*; Star, "Phenomenology of Conventions"; Star and Lampland, "Reckoning with Standards."
80. Thévenot, "Governing Life by Standards," 802.
81. Thévenot, "Governing Life by Standards," 805.
82. O'Connell, "Metrology"; Velkar, "Rethinking Metrology."
83. Velkar, "Rethinking Metrology."
84. Chatterjee, "Women's Question," 238.
85. Uribe, *Las Dinámicas del Tiempo*, 21 and 38.
86. Adib and Emiljanowicz, "Colonial Time in Tension," 1,225.
87. West-Pavlov, *Temporalities*, 157.
88. Dirklik, "Radicalism," 157.
89. Thévenot, "Measure for Measure," 46.
90. White, *Medieval Religion and Technology*, 75.

Chapter 1

Producing Public Timekeeping and Time Pluralism

> Civilization advances by extending the number of important operations which we can perform without thinking about them.[1]
>
> —Alfred North Whitehead, *An Introduction to Mathematics*

The current global standard timescale, UTC, hides debates over time pluralism. The typical clock user does not know who creates, distributes, and manages UTC. But there are elements of UTC that suggest the system is less consistent than lay ideas of science would expect. To my knowledge, time is the only system of measurement where the small units are not directly and uniformly convertible into the large units. When a leap second (see chapters 7 and 8) is added, there is a sixty-one-second minute instead of a sixty-second minute. Gregorian months are not of equal length. A leap year is one day longer than other years. Due to the leap-year formula, centuries that end in a year divisible by four hundred have one more leap year than other centuries. This cobbled-together system inherited from a variety of cultural traditions cannot be described as an inevitable outcome of scientific and technological progress. It makes more sense as a sedimented set of cultural inheritances reinforced by cultural habits.

The time pluralism implied by these awkwardly convertible units of time results from timescale pluralism being common in the past and persisting into the present. Most histories of timekeeping obscure this, however. As a result, the vast majority of works on the history of timekeeping are what I call ancestral histories. They are shaped by the current cultural naturalization of uniform time standards. They emphasize the small group of scientific and technological experts who addressed time-keeping challenges rather than the global debates generated by their decisions.[2] The implicit ending of these historical narratives is the single timescale and the limited number of time-keeping technologies that dominate current practices—Paul Glennie and Nigel Thrift call such narratives "a 'genealogy of the present.'"[3] Often these histories begin by discussing temporal or canonical hours. These divided the daylight period into twelve equal parts. Because the duration of daylight varies with latitude and time of year, the canonical hours were not uniform. Coexisting with these hours were equinoctial hours that divided the day into twenty-four equal parts. These hours were uniform, but more difficult to determine during the daytime than canonical hours. The dominant narrative implies an eventual preference for equinoctial hours. Even then, there was a variation in the length of day throughout the year, but again, by creating "mean time" this variation was transformed into uniform units by averaging the length of day. Rather than mean time being determined for every location, people adopted time zones and eventually there was a "universal time" tied to the prime meridian—known as *Greenwich Mean Time.* With the transition.to atomic timekeeping, this became UTC. The diagram in figure 1.1 represents the common structure of these genealogical narratives of modern timekeeping.

If, for instance, one looks at works such as Landes's *Revolution in Time*[4] or Dohrn-van Rossum's *History of the Hour,*[5] one finds passing references to alternative time-reckoning technologies and techniques, but not much discussion of them. Landes concentrates mostly on the differences between seasonally variable canonical hours and clock hours,[6] whereas Dohrn-van Rossum devotes several pages to different ways of counting hours.[7] In fact, at times, Landes is dismissive of technologies of timekeeping that are not obviously ancestral to European clocks, such as when he describes Su Song's clepsydra in China as "a dead end," and criticizes Joseph Needham's efforts to bring to light Chinese horology as "optative history, the story of what should have been."[8] Even Needham tried to place Chinese horology in an ancestral narrative of European clocks,[9] rather than examining it as a cultural alternative to European timekeeping. In his work *Clocks and Culture,* Cipolla[10] documents Western and Chinese timekeeping, posing an implicit temporal pluralism, but he does not explore the possibility of

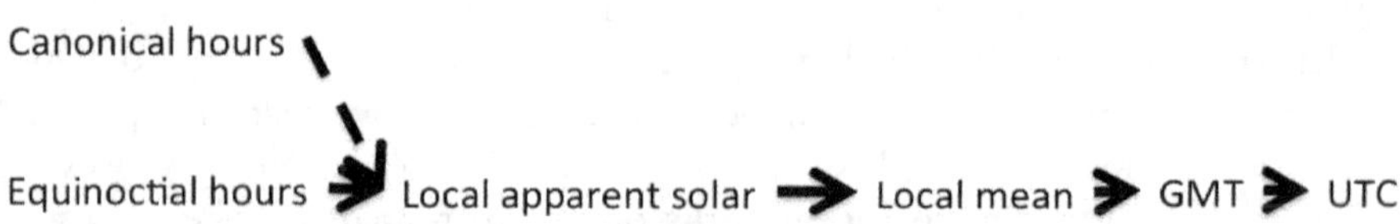

Figure 1.1. Ancestral history of European timescales. © Kevin K. Birth.

different timescales existing in the same cultural context. E. P. Thompson adopted this perspective as well, not only seeing clocks as the obvious replacement of what came before in Europe, but also noting that the dominance of clock time and work discipline "is a problem which the peoples of the developing world must live through and grow through."[11] Akhil Gupta[12] argues that these sorts of narratives not only privilege progress but create a what Johannes Fabian[13] calls "allochronism." This term is based on the classical Greek word for "other," "*allo*." Fabian uses it to emphasize the tendency of ethnography to describe cultural others as inhabiting a time other than that of the ethnographer and the readers of ethnography. One common form of allochronism is the trope of "primitive cultures" that early ethnographies used to represent living societies as if they represented the way "civilized cultures" had been in the past. For both these reasons, Gupta rejects this unilineal logic both in terms of the inevitable outcome Thompson envisioned and in terms of the ability of Western ideas of time to suppress other temporalities. Supporting Gupta's argument, recently there have been a growing number of explorations of how timescale pluralism persisted in the face of colonialism and the globalization of time standards.[14]

Within time metrology, recounting the history of modern timekeeping has become a distinctive genre. The core component of these histories is the transition from astronomical timekeeping to atomic timekeeping and with it the decision-making process that produced UTC as the global time standard and TAI as an important reference timescale for UTC. As these works point out, as timekeeping became more accurate and precise, issues emerged within astronomical timekeeping and the determination of a unit of measure for time.[15] What haunts such histories is the pursuit of reconciling actual clocks with the idea of an ideal clock as a reliable, continuous, uniform giver of time measurement.

These stories of progress are common in studies of technology, but there are problems with them. In his work on technology, Arnold Pacey argues that such narratives tend to "be over-selective, and lead us to overlook the fact that improvements in one dimension are sometimes accompanied by less desirable developments elsewhere."[16] He adds that narratives of progress also tend to be simplifications that omit "irregularities."[17] For most of the history of the idea of the clock, material clocks have been sources of irregularities. For most of their history, they were set to Earth's rotation—itself irregular. Clocks (and the Earth) run fast or slow; clocks need regular maintenance; clocks need to be set; clocks need to be synchronized; and clocks break. The focus on the idea of the clock deflects attention away from the messiness of actual manifestations of clock technology. Moreover, the emphasis on progress towards an imagined ideal clock treats time pluralism as something undesirable.

Efforts to Suppress Time Pluralism for the Sake of the Public

Improvements in the accuracy of clocks led to the recognition that Earth's rotational period was not a good timekeeper. The irregularity of Earth's rotation and the pursuit of idealized regularity in clocks are important elements in the development of time metrology—a field cognizant of the messy relation between clocks and the idea of a uniform, reliable, accurate, and precise timescale, in other words, an ideal clock. Occasionally, efforts to explain time metrology and its practices target the general public, but with limited success. In 1939, the French newspaper *Le Matin* tried to explain the then state of the field to its readers by means of an interview with Armond Lambert, the head of the International Bureau of the Hour (BIH). The reporter asked him, "Quelle heure est-il?" (What time is it?) …

Lambert responded, "Quelle heure ? Mais de quelle heure parlez-vous, monsieur ? Il y a l'heure du Soleil qui est valuable pour les cocoricos villageois, mais qui en dehors des basses-cours est inutilisable." (What time? But of what time do you speak, sir? There is solar time, which is valuable for village cockcrows, but that is unusable outside of the barnyard).

The reporter, not to be fazed by this scientist, then tried again, asking, "Mais l'heure de vos pendules, monsieur?" (But the time on your clocks, sir?).

Lambert answered, "Nos pendules marquent deux temps differrents : l'heure siderale donnée par le astres et l'heure de temps moyen qui se calculi pour donner l'heure universelle …" (Our clocks mark two different times: sidereal time as given by the stars, and mean time as calculated in order to provide universal time …).

Stubbornly trying to pin Lambert down, the reporter tried another approach, asking, "Mais le signaux horaires que vous émettez exacts au centième de seconde? (But the hour signals that you emit that are exact to a hundredth of a second?)

"… Ces signaux ne sont que des prévision horaires." (Those signals are only of the hourly projections.) came the answer from Lambert.[18]

Lambert's answers clearly frustrated yet intrigued the reporter. His answers reflected his good humor. Sadly, during World War II, Lambert was sent to Auschwitz by the Nazis, where he died.[19]

What Lambert attempted to explain was that there are different ways to think about time in relationship to Earth's rotation. Radios meant that time signals could be broadcast, but the relationship between the broadcast signals and the measurement of time would differ from what the public thought. The signals were "projections," according to Lambert. The reporter wanted Lambert to indicate the true time and ideal clock; Lambert wanted to indicate that the broadcast time was a representation and not the "true time," and that the observatory measured two different times: sidereal time and solar mean time. The public wanted the ontological realism of the perfect clock; the best science could offer was a simulacrum using clocks.

As metrological histories make clear, the need to create a time standard that could be replicated with the proper equipment and that was uniform, continuous, and stable is what shaped decision-making processes.[20] These histories, then, while they address multiple timescales, do so only to highlight the choice of one particular timescale. As such, they are greatly insightful, but at the same time, they implicitly suggest that the coexistence of multiple reference timescales is a problem, and that one timescale must be privileged. Yet, there is no reason to assume that the necessary outcome of horological progress had to be a single timescale. Multiple timescales were used at the same time in the past, and as David Rooney points out, even today there are many measures for which there are multiple scales, such as temperature.[21]

The telling of history to privilege one timescale—that associated with the ideal clock—results from the success of the broadcasted timescale and the efforts to avoid confusion by disseminating a single timescale. Like the standard ways of telling the history of timekeeping, the metrological histories do not include all the alternatives. The metrological narratives are better at capturing debates about time than the historical ones, but this is solely within the context of the aftermath of the decision to seek a uniform, continuous timescale. The diversity of timekeeping before that goal emerged in post-Enlightenment Europe is left out of the metrological narrative. If one attempts to represent all the alternative timescales that existed just in early modern Europe, a picture almost too complex to generate a simple narrative emerges (see figure 1.2). In this complex picture, it must be kept in mind that Jewish, Muslim, Hindu, Chinese, and Japanese timescales were still different from the Christian European timescales. A total global picture of the diversity of timescales is even more complicated than the one presented in figure 1.2.

Figure 1.2 indicates that timescale pluralism used to be the norm in Europe. This is not a simple matter, however. The names of the different timescales imply regional methods of timekeeping, not that these different methods necessarily coexisted in the same place. Astronomical hours were a count of twenty-four hours of equal length beginning at solar noon. These were associated with astronomers, particularly Ptolemy, because solar noon, or when the Sun transited one's meridian, was a clear and obvious daily temporal landmark. The term "Greek hours" applies to a count of twenty-four hours that began at midnight. Small/French hours were two counts of twelve hours each that began at midnight and noon, respectively, and got their name because of their association with small clocks. The great or Nuremberg hours were hours of equal length but applied to different counts for the daylight and nighttime periods. As a result, at the summer solstice, the daytime hours could go as high as eighteen o'clock followed by one o'clock at night. The Italian hours were a count of twenty-four equal hours that began with the ringing of the evening Angelus bell in churches, which typically happened about thirty minutes after sunset. Babylonian hours were a count of hours that began at sunrise.

Many famous surviving clocks from early modern Europe are examples of such timescale pluralism—a cosmopolitanism in timekeeping that sharply contrasts with current logics. Such surviving clocks remind us that clock time could have been different from what it is at present, and that clock time in the future might be different from what it is now. These clocks also point to another feature that is not prominent in current conceptualizations of time—the ability to convert between different timescales. For instance, the famous Astronomical Clock of Prague allows the easy conversion between Italian/Bohemian hours, small hours, and canonical hours.

The prominent display of Bohemian/Italian hours on the Prague clock indicates their importance. Dohrn-van Rossum viewed them as the oldest form of mechanical equinoctial hours.[22] These hours are well suited for anticipating dusk, but not so well suited for anticipating dawn. Jérôme Lalande, who traveled to Italy in the 1760s, wrote that while foreigners find Italian hours strange because twelve o'clock shifts during the year,

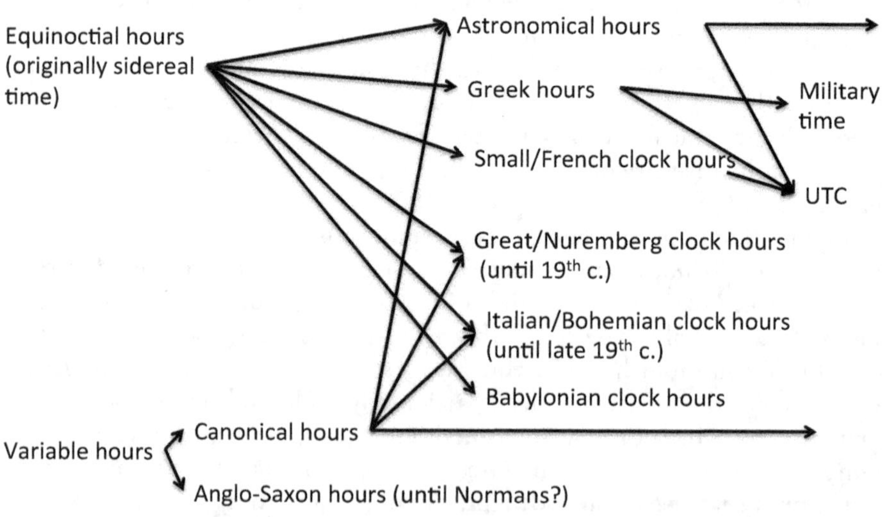

Figure 1.2. An over-simplified genealogy of European timescales. © Kevin K. Birth.

> …si l'on examine la chose sans prejudge, l'on trouvera que la method italienne es peut-être la plus naturelle ; car avant qu'on eût trouvé des machines pour mesurer le temps on ne pouvoit partir d'un point plus sensible à tous les yeux, que la chute du jour, & finir ses occupations plus naturellement que par la fin de la lumière. (…if one examines this thing without prejudging it, then one finds that the Italian method is perhaps the most natural, because before one had invented machines to measure time one could not start from a point more sensitive to all eyes than the end of the day, and more naturally to finish work than by the end of light.)[23]

In 1786, Johann Wolfgang von Goethe struggled to relate Italian hours to the hour system familiar to him.[24] Another period traveler, Christopher Hervey, wrote that the Italian method of reckoning hours had advantages, "and to a traveler in particular, as you always know the exact minute of the Sun's setting from your watch, and, consequently, how much or little day-light remains, be the weather ever so misty, to complete your journey and get time enough for the gates of foreign towns which generally are shut at that time."[25] Hervey alludes, then, to the usefulness of such hours—the congruence between protecting public safety and the hour system adopted. This timescale was not an abstract, general measure of time, but a time anchored to an important moment, the closing of the city gates, in a way that was useful to residents and travelers.

Viewing clocks as tools of public policy shifts the question about the Prague clock's measurement of time to its policy implications—it is a cosmopolitan device that allowed the public to convert between the Church's dominant hour system, the hour system used for astronomy and astrology, and Prague's hour system. It points to a time and culture in which there was civic pride in displaying multiple timescales.

In figures 1.3 and 1.4 are two drawings derived from photographs of the relevant dials of the Prague clock at twenty o'clock in Bohemian hours at two different times of year. The drawings remove the ecliptic ring, on which the image of the Sun travels, for an easier view of the dials and to allow the number of the different hour systems to be seen with greater clarity than in a photograph. In figure 1.3, twenty o'clock in Bohemian time is also about two o'clock in small hours (our present system), and the location of the Sun icon indicates that it is between the eighth and ninth hours in the seasonally variable canonical hours. In contrast, in figure 1.4, while it is still twenty o'clock in Bohemian hours, since it is a different time of year, it is about 4:00 PM in small hours, and about the ninth hour in canonical reckoning.

The Prague clock's display of timescale pluralism indicates the cosmopolitan nature of Prague when the clock was built. Under the Holy Roman Emperor, Charles IV, this city became a major trade center that would propel it into becoming one of the most important cities in Europe. Its stature as a meeting place of different religious and cultural traditions attracted another Holy Roman Emperor, Rudolf II, to make it his capital, and under him and his patronage of astronomy, horology, alchemy, and art, it became the most cosmopolitan city and preeminent center of learning in Europe at the time.[26]

The consciousness of clock time begins relatively early in Prague. The late fourteenth-century regulations issued by Archbishop Arnestus of Prague for the *mansionarii*, a type of chaplain, emphasized the keeping of the Divine Office at the correct times according to canonical hours. These regulations even include a reference to the use of a horologium—whether this is a clepsydra or a mechanical clock is not clear.[27] The use of clock time to indicate the exact moment of a significant event appears in chronologies that start in the early fifteenth century—shortly after the famous clock was built. The emphasis on recording the exact time hints at the possibility of there being an astrological significance to these times since such precision during the period was most often found in association with astrological charts. In the *Chronicon* compiled by Gelasius Dobner, the earliest reference to a clock time had to do with the events related to the capturing of Prague Castle in 1420 on "*VI. Die mensis Aprilis hora XVI. Horologii.*"[28] After this first use, Dobner records other events using clock time for 1427 and 1437.

The ability of Prague's Old Town Hall Astronomical Clock to convert between canonical, Bohemian, and small clock hours fits with an urban environment where religious communities still organized their day according to

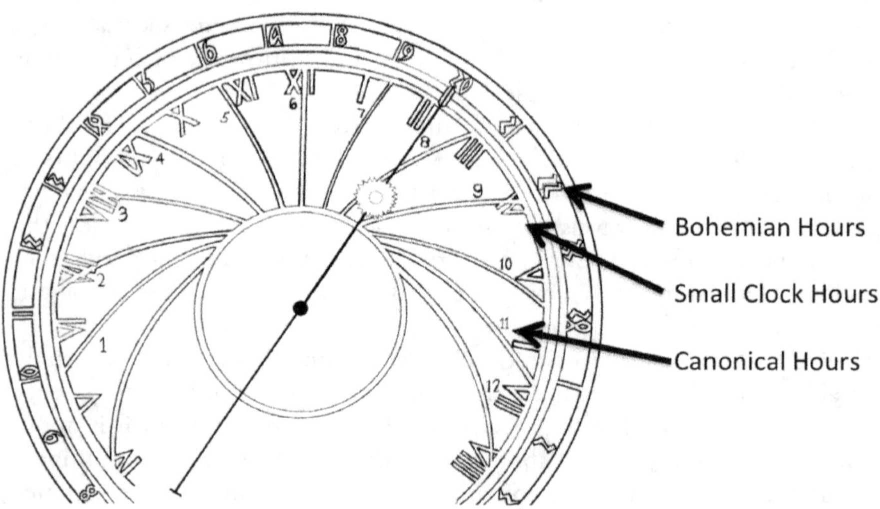

Figure 1.3. Timescale conversion in the Astronomical Clock of Prague—twenty o'clock, 2 PM, between the eighth and ninth canonical hours around the equinox. © Kevin K. Birth.

the canonical hours, and trade was flourishing. The astrological information indicated by the clock also fits with a cultural context obsessed with the planetary forces acting upon any given moment, and with how such forces were part of the interpretive framework of significant events. The temporal pluralism that existed in this context was, therefore, an indication of the different uses of timescales, and a view of timescales as special-purpose tools rather than simulacra of a standardized, hegemonic ideal clock. Uniform timekeeping was not appropriate for religious purposes, nor did it capture the kairotic nature of astrologically significant moments. Imposing a single timescale also conflicted with the sensibilities of travelers who were familiar with other timescales. The cosmopolitan nature of the clock both in terms of how it indicated multiple timescales and of how these multiple timescales were employed in different systems of generating meaning about time, fit the confluence of religious, economic, and scientific interests in the city—a confluence the city's populace seemingly viewed as unique. In other words, it is not possible to understand the clock without understanding the social relations it served.

Prague was not alone. Nuremberg prided itself on its own way of counting hours, and surviving broadsides printed there provide a means of converting between multiple timescales.[29]

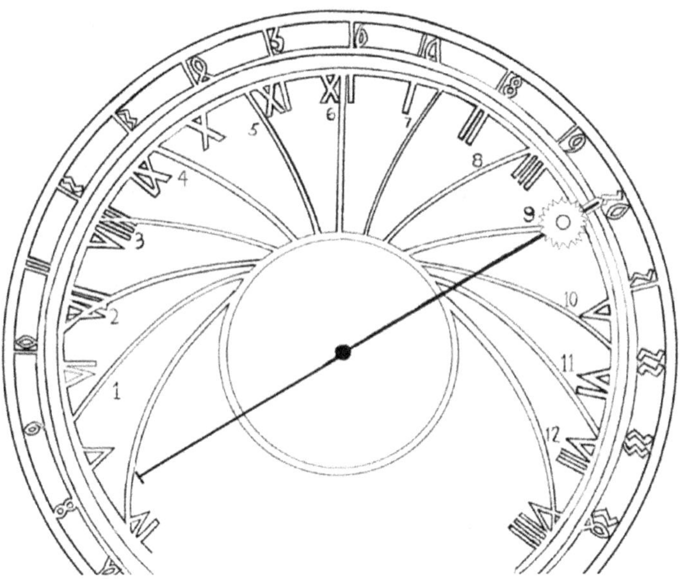

Figure 1.4. Timescale conversion in the Astronomical Clock of Prague— twenty o'clock, 4 PM, and the ninth canonical hour around late May or early August. © Kevin K. Birth.

One surviving broadside from Nuremberg in 1589, entitled "Vergleichung nachvolgender Siben Uhren," is a testament to timescale pluralism. It lists the following: Greek hours (1–24 starting at midnight), astronomical hours (1–24 starting when the Sun crosses the meridian), small hours (two series of 1–12 starting at midnight and noon respectively), Nuremberg hours (two separate counts of daylight and nighttime hours), foreign hours (also known as Babylonian hours, numbered 1–24 starting at sunrise), and Bohemian hours (1–24 starting at sunset).[30] In another surviving broadside[31] from Nuremberg—this one printed in 1564—canonical hours are represented. Each of the nine columns represents times of year with roughly equivalent amounts of daylight. For instance, the period of the spring and fall equinoxes shares the same column. Each column is then divided into three subcolumns. On the right is the subcolumn for canonical hours—the space between these hours changes throughout the year. The middle subcolumn contains the small hours, and the left subcolumn contains Nuremberg hours in which one o'clock in the day is always one hour after sunrise and one o'clock at night is always one hour after sunset.

In writing about the late Middle Ages, Lynn White described the town clocks that displayed such time cosmopolitanism as sources of "civic pride"[32]—a contrast to our age when the stability, precision, and accuracy of a suite of atomic clocks to indicate the global standard, UTC, is a source of pride. Maybe the social significance of the Prague clock in defining the city's prestige is why the legend grew that the maker of the clock was blinded after finishing it so that he could not make a similar clock for another city—the clock was not just an indicator of Prague's special place in Europe, but a means of ensuring that special place. That the legend is not true does not diminish its cultural significance. The cultural representations of the past that are salient often differ from the careful construction of scholarly history.[33] The legend of the blinding of the clockmaker for Prague need not be factual to have an impact like the stories of King Arthur, or of Columbus discovering America. It is the affect-laden meanings of these tales that matter, not how well they are corroborated by careful historical research. Moreover, such tales become reinforced by artifacts or sites in the present that serve as indices of the events as told in the stories. In many ways, the Astronomical Clock of Prague is more tangible than the remnants of Columbus or King Arthur. The story of the clockmaker's blinding also parallels current time policy. Those involved in designing and managing the state-of-the-art atomic clocks keep cutting-edge technology a secret.

The constellation of science, technology, and secrecy has remained a part of clock development, but there has been a shift away from time pluralism and cosmopolitanism reflected by conversion between multiple hour systems found on the Prague clock to time standardization and the singular ideal clock. The result of this shift—the current emphasis on uniformity—fosters the tendency only to see uniformity when looking at the past. As a result, many ideas and uses of time-keeping technologies of the past become doubly obscured—first by the secrecy of those who developed those technologies and second by the unawareness in the present of those technologies. All clocks start to look alike, and it is not possible to grasp the stories different clocks tell.

Clocks themselves obscure technical knowledge. The modern clock represents time in a way that young children can read,[34] and since most people's education about timekeeping ends at a young age, advanced clock technology and time science are safely kept from adult minds. It is kept secret in another way. Telling time with the accuracy and precision currently desired is too complex for a single human mind—it takes a cohort of technicians, engineers, and scientists distributed across the globe. Despite this complexity, the childlike knowledge most people bring to timekeeping is sufficient to use the representations the technology produces. It is an example of the limits of information technology not simply being a result of the complexity of the machine, but in the attention humans can give to what the machine produces.[35] The politicians who define policies in relationship to time currently operate with childlike naivete about time metrology. They accept the image of the ideal clock and assume that clocks are sufficiently synchronized and perform sufficiently well for regulations defined in reference to clock time to be followed and enforced. This, too, lends itself to a distorted view of the past, specifically, a distorted view of the nexus of timekeeping and power. The relationship of power to clocks was as variable as the clocks themselves, and often power was manifest not in the control over an abstract standard indicated by clocks, but in the control over the specific clock that set the standard. The next several chapters explore varieties of this relationship between power, time pluralism, and clocks. In late medieval York, the power to regulate markets depended on the power to control specific clocks and their bells. In early modern Habsburg domains of Europe, cosmopolitan time pluralism indicated the reach and diversity of Habsburg power. But as chapter 4 discusses, such cosmopolitanism creates a need for scientists to make choices about which timescale to use to record observations, and these choices had implications for how the globe was imagined.

Notes

1. Whitehead, *Introduction to Mathematics*, 46.
2. Ogle, *Global Transformation of Time*, 11–17.
3. Glennie and Thrift, *Shaping the Day*, 27.
4. Landes, *Revolution in Time*.
5. Dohrn-van Rossum, *History of the Hour*.
6. Landes, *Revolution in Time*, 76–78.
7. Dohrn-van Rossum, *History of the Hour*, 113–17.
8. Landes, *Revolution in Time*, 21.
9. Needham, *Science and Civilisation in China, volume 3*, 315–19, 359–66; Needham, *Science and Civilization in China, volume 4*, 435–534; Needham, Ling, and de Solla Price, *Heavenly Clockwork*.
10. Cipolla, *Clocks and Culture*.
11. Thompson, "Industrial Capitalism," 95.
12. Gupta, "Reincarnation of Souls."
13. Fabian, *Time and the Other*.
14. Ogle, *Global Transformation of Time*; Barak, *On Time*; Frumer, *Making Time*; Wischnitzer, *Reading Clocks*.
15. McCarthy and Seidelmann, *Time*; Audoin and Guinot, *The Measurement of Time*.
16. Pacey, *Culture of Technology*, 14.
17. Pacey, *Culture of Technology*, 14.
18. *Le Matin*, "Quelle heure est-il?" 2.
19. Charvin. "Allocutions," 27–28.
20. Audoin and Guinot, *Measurement of Time*; Levine, "History of Time and Frequency"; McCarthy and Seidelmann, *Time*; Winkler, "Developments."
21. Rooney, *About Time*, 248.
22. Dohrn-van Rossum, "Uhren, Glocken und Zeitorganisation," 67.
23. Lalande, *Voyage d'un François en Italie*, xxxiii.
24. Goethe, *Italian Journey*, Verona, Sept. 17.
25. Hervey, *Letters from Portugal*, 402.
26. Evans, *Rudolf II and his World*, 74.
27. Dobner, *Monumenta historica Boemiae*, 319–20.
28. Dobner, *Monumenta historica Boemiae*, 59.
29. See Gaab, "Die grosse Nürnbergische Uhr," 50–60.
30. Ammann, *Vergleichung*. Dackerman, *Prints and the Pursuit*, 75–76.
31. R. F., *Kalender*.
32. White, *Medieval Technology and Social Change*, 124.
33. Birth, "Immanent Past," and "Past Times"; Trouillot, *Silencing the Past*.
34. Birth, *Time Blind*, x–xi.
35. Lanham, *Electronic Word*, 227.

Chapter 2

A Bell for Every Market

Time Pluralism in Late Medieval York

> Our quest for the dimension of time will therefore be neither harmonious nor linear. Its advance will be marked by a polyphony of dramatic shocks.[1]
>
> —Édouard Glissant, *Caribbean Discourse*

Before the modern consolidation of metrological institutions and the seeking of a single publicly available reference timescale, people were subject to multiple sources of time that did not agree. Their choices of which sources to heed depended on the relationship between agency and power. To grasp this involves looking at the daily experience of timekeeping in the period before the emergence of centralized time-giving institutions.

Medieval Europe defined and regulated activity using bells as time signals. With evidence about bells, one can identify institutions that defined their own timescales, and from that reconstruct the distribution and use of different timescales within a social space such as a city. In this period there was no singular concept of a clock, no synchronization, and a greater concern for institutional autonomy than for uniform standards. That noted, the knowledge that there were multiple bells governing different aspects of urban life does not begin to explore the possibilities of how individuals pitted those signals against each other or used sonic differences for individual or collective advantage. Bells associated with civic regulation targeted public spaces, yet the nature of sound is that their bell peals would penetrate private spaces and transcend civic boundaries.

The fifteenth-century book *Dives et Pauper*, attributed to the Carmelite Henry Parker, states: "in citees and townes men rule them by the cloke /

& yit properly to speke/ the clocke ruleth natt them but a man ruleth the clock."[2] In this context, "clocke" refers to a bell that indicates time, and not necessarily to a mechanical device. The author's point was that a clock's time was "ruled" by a person, rather than by a generic concept of every clock being synchronized to a single time source.

The importance of bells as givers of time contradicts the frequent portrayal of medieval Europeans as paying little attention to time that is found in the work of Marc Bloch,[3] G. J. Whitrow,[4] Paolo Rossi,[5] Alfred Crosby,[6] and Lucien Febvre.[7] In contrast, Jacques Le Goff pointed out that, once we accept that the reckoning of time in the Middle Ages was different than it is now, "far from being indifferent to time, men in the Middle Ages were singularly sensitive to it."[8] In addition, there is evidence of the representation of work in terms of units of time, and the docking of workers' wages for missing work time predates the use of the clock to represent labor.[9]

Even after clocks became available there were multiple ways of reckoning time.[10] For a long period, temporal/canonical hours rung by bell ringers were more important than clock-defined equinoctial hours. This persisted into some parts of France until the early twentieth century.[11] In Florence, the main time-givers were the bells of the Palazzo Vecchio, which were rung according to canonical hours by bell ringers, yet this building also had a clock by the end of the fourteenth century that tolled the hours. This clock's bell "represented the uncoupling of timekeeping from the day's other integrated secular and sacred temporal rhythms" and consequently, was not mentioned in the 1415 statutes on bell ringing.[12]

By the end of the fourteenth century, throughout much of Europe a new time consciousness emerged as indicated by references to time in city ordinances.[13] Along with the appearance of time-keeping language in statutes was a subtle shift in how time is discussed, so that events were specified as happening at a particular "clock hour."[14] In contrast to canonical hours, clock hours were linguistically marked with modifiers such as being "of the clock," and they remain so in many languages, such as English in which the modifier "o'clock" is still commonly used. This contrasts with canonical hours, which were not linguistically marked by any modifying term like "o'clock"—they were Matins, Lauds, Prime, Terce, Sext, Nones, Vespers, and Compline. Clocks were not synchronized. The medieval city involved many time signals—some for the entire populace, some for certain activities or groups such as monks.[15] Peter Burke observes that in Europe, from the medieval through the early modern period, "awareness of alternative systems of time reckoning was part of everyday life."[16]

This chapter examines the multiple times signaled by bells in late medieval York. The surviving record from York displays a diversity of time-related regulations that specify their sources of time. While the history of clocks often focuses on monasteries, David Rooney notes that many of the earliest references to mechanical clocks are associated with market regulations in cities.[17] This makes it possible to match regulations and time signals to particular political and religious institutions within the city, thereby giving a picture of how these institutions sought to influence social rhythms. The use of time to regulate markets in order to give advantages to particular groups of merchants seems a widespread practice in the late Middle Ages and has been documented for Liège[18] and Bristol.[19] In York, even after the arrival of public clocks, residents used several ways of indicating time. A uniform system of timekeeping was not an 'aspirational ideal' during the late medieval period in the city.[20] In effect, direct control over timekeeping guided regulations, rather than regulations referring to a shared concept of synchronized clock time. Court cases corroborate the coexistence of multiple ways of reckoning and signaling time. There was no established convention for using equinoctial hours, temporal hours, the position of the Sun, or references to dusk or dawn.[21]

In York, archaeologists recovered one of the earliest artifacts yet discovered related to clocks. It is the seal of Robert the Horologer found in an archaeological stratum dating to around 1300.[22] Either Robert ranged widely, or he was not an isolated craftsman: in 1272, Henry III"s building accounts list Robert of Beverly (a town in Yorkshire) as supervising the construction of a clock at the church of Westminster.[23] Aside from the age of Robert's seal, its portrayal of a wheel divided by eight lines is an indication that the time associated with early clocks may have been different than the system to which we are accustomed. The division of a clock face by eight lines is not unusual, however. It is associated with Norse time reckoning,[24] Anglo-Saxon sundials,[25] and also to the practice of reckoning equinoctial hours by means of the compass rose.[26] This sort of reckoning was standard for astronomical observations and for anticipating tides. The Northumbrian monk Bede is viewed as coming up with the first tidal tables, and there are several extant tidal tables and diagrams from the period between Bede's death in 735 and the advent of clocks in the thirteenth century.[27] In the period of Robert the Horologer the conventional way of reckoning time was temporal/canonical hours that varied in length with the season.

Variation in the determination of temporal/canonical hours and the emergence of clocks indicating equinoctial hours contributed to a lack of synchronized time signals. While it is difficult to capture this temporal cacophony in its entirety, surviving regulations provide a partial picture. Such regulations address the lack of synchronization by specifying the source of time for the regulation. The opening of a market was not merely at, say, "seven of the clock," but at "seven of the clock" of a specific time source. The multiple sources of time created a complex timescape, a term coined by Barbara Adam to mean contextual constellation of "rhythmicities, their timings and tempos, their changes and contingencies."[28] Medieval York had a complex timescape of different time-givers tied to different institutions within the city.

Not only was synchronization between time sources was not a goal, but it was antithetical to many political aims. To synchronize to one time-giver would be to privilege it above others, thereby privileging the institution that controlled that time-giver. Cacophony indicated the assertion of equality. Control over time has always been political, but what makes the late medieval European timescape different from the modern age is that it allowed for the proliferation of different time standards as sources of time, rather than the current practice which recognizes a single standard and proliferates references to that standard.

By the seventeenth century, references to specific clocks and bells disappear from York's regulations. This indicates the emergence of time standards agreed upon by convention even if they had not yet been enshrined in law. The emerging agreement about time in the seventeenth century also probably reflects two simultaneous trends in York: the consolidation of civic control under the city government that occurred after the Reformation as Henry VIII destroyed abbeys' power and improvements in local horological craft.

Bells and Sociality

The first clocks were automated bell ringers. Bells were the key signal in organizing and regulating civic and religious time. Niall Atkinson observes, "Sound was a vital element within an urban information system" and "it formed part of people's way of navigating in time, space and in the social world of the city."[29] Based on his study of Renaissance Florence, he argues that "Bells were arguably the most important urban sounds in the pre-modern world since they connected inhabitants to the most intimate aspects of their lives: to their religion, their spiritual expression, their past, their dead, their government, their safety, their civic duties, their labor and their collective celebrations."[30] Clocks were part of a set of institutions and practices associated with bells that ranged from bell towers to town criers carrying hand bells, to mechanical clocks. Bells served to create an "auditory community."[31] But hand-rung bells were more useful than clocks. Clocks could only indicate the time, and not signal the other messages that bells conveyed, such as deaths, important civic events, and emergencies.

The diversity of bells associated with the variety of social institutions also indicated various roles and statuses within the city. The bells one heeded indicated one's social, religious, economic, and political networks. The bells governing market activities often differed from those associated with daily liturgies. Dohrn-van Rossum notes that the multiplicity of bells led to disputes over which institutions had the right to ring their bells. Since there were different uses and patterns of ringing bells, there was "hardly a day resembling another."[32]

Bell peals varied according to the message conveyed. The ringing of bells when somebody died was different from the ringing used to indicate time. To be a bell ringer meant more than just pulling on a rope. Different purposes had different peals, different paces, different rhythms, and in towers with multiple bells, different sequences. For instance, in discussing the bells and bell ringers of York Minster, David Potter quotes a thirteenth-century statute that indicated "that before matins the ringing was to last for as long as it took to say 'the matins of Our Lady'; before prime it was to last long enough for someone to come to the cathedral from three miles away, and before other services it was to last long enough for the canons to come to the minster from their houses."[33]

In York, bells were displays of authority,[34] and bell founders respected. From 1206 to 1211, Gerard the bell-founder was the prepositus of the city—before the office of the mayor was established, the prepositus was the senior city official who answered to the county sheriff.[35] An early fourteenth-century bell founder, Richard Tunnoc, became mayor and later a Member of Parliament. Tunnoc sponsored the bell founders' stained-glass window in York Minster. A foundry located near York Minster made at least small bells.[36] While no casting pits for large bells have been found in York, the bells portrayed in the bell founders' window in York Minster and the widespread distribution of bells documented by George Benson,[37] who found 450 surviving bells cast in York between the fourteenth and eighteenth centuries, indicate that large bells were produced somewhere in the city.

The multitude of bells distributed across the medieval city indicated the different jurisdictions and "liberties" (i.e., institutions not subject to the city's civic authorities but having their own governance). In York during the fourteenth and fifteenth centuries, several distinct political interests were competing for influence. The largest and loudest bells were also the bells most mentioned in regulations. These bells were located in four structures representing four different sources of power within the city: the Cathedral Church of St. Peter, more colloquially known as York Minster; St. William's Chapel, which was part of the complex of buildings that were the seat of the civic corporation's power; All Saints' Church in Pavement, a church dominated by the merchant guilds; and St. Mary's Abbey, located just outside of the city to the west, and one of the wealthiest and most powerful monasteries in England. York Minster and St. Mary's Abbey were politically independent from the city and often disputed claims to property with the city's government, also known as the city's civic corporation. The powerful guilds and the civic corporation played complementary roles, with the merchant guilds regulating their markets, and craft guilds their quality of products and labor practices. The city government and its ward courts enforced criminal law, maintained the gates to the city, regulated the times of their opening, and heard civil cases that were not under the jurisdiction of a particular guild. About the different civic institutions, Sarah Rees Jones and David Palliser write: "Within the city's jurisdiction, but wealthy and powerful enough to overlap with rather than be subordinate, there developed a few large social and religious guilds and fraternities."[38]

Truthfully, I portray an overly simplistic picture. Palliser found that "by 1300 there were at least 14 islands of jurisdiction within or just outside the walls where the bailiffs could not make arrests and where men and women could work without taking up the city's freedom."[39] The result was that the "regulation of the city and its residents was the result of many compromises made between the royal law, the civic magistrates' policies and the customs and practices of the 'commons' of the city."[40] It is not clear how many of these different jurisdictions used their bells to signal their own times, but it is clear that the guilds, the civic corporation, St. Mary's Abbey, and York Minster were particularly important time-givers for purposes of organizing social and economic activity, and that their time signals were distinct and not synchronized. This created an urban soundscape in which residents had to contend with distinguishing different bells and to recognize the association of those bells with different powers. Such a situation was not unusual, for the nature of bells was that their sound carried beyond the area to which it belonged.[41] Sometimes these different institutions would be in conflict, but there is also evidence of some cooperation—this is particularly the case when regulations refer to time signals from different institutions. There is also a sense that, with no one institution having absolute dominance over the cycles of activity in York, there were gaps in the regulations that entrepreneurial spirits, particularly those interested in victualing and innkeeping, could exploit.

Bells for Guilds

The guilds of York had considerable control over labor and commerce. The craft guilds drafted their own rules regulating apprenticeships, admission to being a master, quality of products, and work rules. The city council ratified these rules, but this does not indicate direct control over the guilds by the council, only that the council recognized the rules drawn up by the guilds.

References to time can be found in the ordinances that the city approved. In fact, the earliest surviving guild ordinance is from 1307. It prohibits girdlers working at night.[42] Such regulations became common. They strove to ensure the quality of products and to ensure quiet for neighborhoods of craftspeople.[43] This seems to affirm Le Goff's assertion that even in the crafts, the ideas of time were dependent on "natural time."[44]

"Natural" time is not straightforward, however. The boundary between day and night is blurred by the period of twilight. The thirteenth-century English Franciscan Bartholomaeus Anglicus described the transition between day and night with a complex interweaving of different time cues that alludes to the inability to use just one indicator to define the beginning of night. These cues included the closing of town gates, the growing length of shadows, the closing of flowers, the decline of heat, the closing of one's pores, the way some animals go to sleep, the way other animals like owls come out, and the routine of herders gathering their flocks.[45] This is a multisensory approach to distinguishing day from twilight, and he sustains this approach when he discusses night, a period that is associated not only with darkness, but also with how "occultat enim rerum colores" (it hides the colors of things). Night is also "frigida et humida" (cold and moist). During night, animals and people sleep, and the stars shine brightly, which allows for discerning the passage of the night.[46] As a result, any regulation that forbade work at night begged the question of how to determine the beginning of night.

Sufficient light for one activity is not sufficient for another. In addition, terrain and buildings affect how shadows are cast, which makes for very different light conditions in nearby locations. There are indications that some crafts attempted to work by candlelight. For instance, there is a regulation of the founders from the late fourteenth century that reads: "que nul du dit artifice overa par lumer del chaundel mais soulement tanqez le lumer de jour lui purra resonablement suffier a overer" (that none of the said workers work by the light of the candle but only until the light of day is able reasonably to suffice to work).[47] So "natural" time is a culturally based determination. Even if there are regulations about the end of the workday in relationship to light, masters probably varied in the extent to which they required those in their employ to work under deteriorating light conditions in the evening.

In fact, in stained glass there is circumstantial evidence of working under changing light conditions into the evening and night. This evidence is in the form of the highly variable darkness of red one observes in some stained-glass windows from the medieval period, without seeing similar variations in darkness in other colors. Likely, this is due to a feature of the eye called the Purkinje effect. This is named after the pioneering scientist Jan Evangelista Purkyně, who wrote:

> Take some colors before daybreak when it begins slowly to get lighter. Initially, one sees only black and grey. Then the brightest colors, red and green, appear darkest. Yellow cannot be distinguished from a rosy red. Blue looks to me to be the most noticeable. Nuances of red, which otherwise burn brightest in daylight, namely carmine, cinnabar, and orange show themselves as the darkest, in contrast to their average brightness. Green appears more bluish, and its yellow tint develops with the increasing daylight.[48]

In dim light, the human eye relies on the rods rather than the cones to see. Rods are sensitive to light, but their color perception is primarily limited to the green-blue wavelengths. Red is an odd color in human visual perception because, regardless of its hue in full light, in dim light it becomes uniformly black. As a result, a glazier working in dim light would be unable to distinguish light red from dark red but would be better able to distinguish differences in shade in yellows, blues, and greens. In York, variation in the shade of red is found in almost all the medieval stained glass in the city, although it is particularly easy to see in the stained glass of Holy Trinity in Goodramgate because of the proximity of the windows to ground level and the large pieces of red glass.

The merchant guilds had a great deal of control over regulating trade. The use of time to regulate markets is ancient and continues in our time, so the application of regulations by means of time in the late medieval period in York comes as no surprise, and Edward Miller describes such regulations as "a constant pre-occupation of the city authorities."[49] Every public market in York was regulated in this way—whether it was the food markets, the textile markets, or the fish markets. The point was to control trade,[50] and, in fact, many regulations in York related to guilds were for the temporal control of markets. The records that survive from guilds primarily depict the administrative desires of the guild authorities,[51] which were often an extension of the desires of civic authorities.[52] They are regulations applied to daily life and not portrayals of daily life.

The greatest number of regulations apply to the selling of food. Food has a complex history within York. York sits in the middle of a fertile agricultural region, but York was also a highly socially stratified city with multiple interests.

One of the most important food markets in York was the Pavement Market, where grain was sold. In the late medieval period, the bell of the nearby All Saints Church regulated the opening of this market. According to post-Reformation church inventories, this church had one of the largest bells in the city, weighing 8,000 pounds.[53] The market itself was probably founded in the tenth to eleventh centuries.[54] The Domesday Book of 1086 indicates the existence of this market before the Norman conquest.[55]

The record of the regulation of the Pavement Market dates back to the late fourteenth century. From that time one encounters a regulation that prohibited the sale of goods in the Pavement Market until "primam pulsacionem ad matutinas in ecclesiis parochialibus" (the first striking of the morning bell in the parochial church).[56] About a century later, one finds a similar regulation that states that bakers cannot enter the Pavement Market: "be fore xij be fully strekyn of the clok of Halous of the Payment."[57] This regulation ensured that bakers did not buy up available stocks before ordinary people seeking to buy food, thereby driving up prices and diminishing supply. Likewise, in 1519: "that the cornebell of the Payment shalbe rong at x of the clok of the market days, and moreover that no baker foryns shal by eny corne ther unto the houre of one at clok at the after nowne the same day."[58]

The regulation of the Pavement Market is the best documented of all the markets associated with guilds, but there are other markets that make an appearance in surviving records as well, and like the Pavement Market, a specific bell regulates these other markets. For instance, in 1518 the fish market opened at "viij of the clock in the mornyng on the market dayes" according to "the bell of Fossebregg, called the Scaytbell."[59]

By 1505, the bell of St. William's Chapel signaled the opening of Pavement Market.[60] This change suggests a consolidation of regulation in the hands of the civic corporation, since this bell also regulated the next market discussed, a market always under the city's control rather than guild control.

Civic Bells

There was time pluralism in York's markets because they were controlled by different institutions, each with its own bells. Thursday Market was a food market regulated by the civic government. It is first mentioned in 1239 in association with the royal larderer, who placed tolls on food sold in the city until 1253.[61] Traders from outside the city who had not achieved the status of freemen of the city sold in the Thursday Market, and freemen of the city could not purchase goods in Thursday Market until several hours after the market opened. As with Pavement Market, this was to control prices and prevent hoarding.[62]

In contrast to the Pavement Market, which opened at Matins, the Thursday Market during the same period did not permit the selling of food before "that sex by stryken open the common bell at Ousebrigg."[63] It is not clear whether "sex" means the canonical hour or a clock hour. The canonical hour would have been at solar noon, which would have had the market opening considerably later than the one at Pavement. If "sex" here might mean 6:00 AM, this gave the Pavement Market a slight edge in its opening time. Matins, the morning liturgical office that determined the timing of the opening of Pavement's market, occurs before dawn. Its beginning is not tied to clock time but is based on a judgment. Moreover, the length of the liturgy associated with Matins expands between the fall and spring equinoxes when the Sun rises later. So even during the winter months when the Sun rises later than 6:00 AM, the beginning of Matins would still occur before sext. Consequently, Pavement Market would always open before Thursday Market. This edge might have been the result of the tension between local food producers and the civic government over the creation of the Thursday Market. Thursday Market attracted traders from outside of the city, which created competition for sellers who were residents of the city.[64]

The common bell at Ouse Bridge that regulated Thursday Market was in St. William's Chapel. Chapels were not churches, but either subordinate to a church or independent of a parish. Secular institutions often established chapels to assert independence from parishes.[65] St. William's Chapel was dedicated to Archbishop William Fitzherbert, York's patron saint. His hagiography attributes to him the miracle of nobody perishing when the bridge across the River Ouse collapsed shortly after his ascendancy to the position of Archbishop of York. Shortly thereafter, William died under circumstances that some viewed as a murder—making him a martyr. At some point after a stone bridge was built across the Ouse, a chapel was constructed on the bridge. The chapel was connected to the Lord Mayor's chambers and associated with the Civic Corporation.[66] Among its features were not only a clock, but also an altar dedicated to St. Eligius, the patron saint of goldsmiths and clockmakers.

During the late medieval period, the bell of St. William's chapel determined the opening of the Thursday Market. This market was over three-hundred meters away from the chapel, but it was also at about the same relative elevation. Between the chapel and the market were slopes down to the river. The acoustics of a clock tower made of wood with slats that only opened toward the market likely overcame this distance. The clock tower's design adopted ideas of resonance and reflecting sound discussed by Vitruvius in book V of his *Ten Books on Architecture*,[67] which continued to be used until Isaac Newton developed a mathematical model of sound waves in the eighteenth century.[68] These ideas are discussed in detail by Athanasius Kircher in his *Phonurgia Nova sive Coniugium Mechanico-Paranympha artis et naturae Phonosophia concinnatum*, where he described architecture for the reflection of sound (phonocamptics). There he commented on how a square structure can amplify sound.[69] Kircher merely developed the metaphor Boethius used in his discussion of the movement of sound. In *Fundamentals of Music,*[70] Boethius suggested that sound moves like a ripple after one tosses a stone in water. Because of Boethius's influence on medieval thought, other works on sound and music repeat this metaphor[71]—most famously Adelard of Bath's dialogue with his nephew.[72] The chapel's location and design of the bell tower likely overcame the three-hundred-meter distance of the market from its time signal.

Regulations concerning the times at which markets could open continued to refer to specific clocks or bells through the mid-sixteenth century, but in 1554, the regulation of Thursday Market referred to time as kept by the Minster clock.[73] This clock was much closer to Thursday Market than St. William's Chapel. It is possible that the clue lies in a 1556 entry in the civic records about the repair of the Ouse Bridge clock.[74] This also implies that the bells of both York Minster and St. William's Chapel could be heard in Thursday Market. Therefore, there must have been some way to distinguish these bells for the market to function.

By the late sixteenth century, both Thursday Market and Pavement Market had structures called *market crosses.* These were buildings on which a sundial was mounted, and in which standards of measurement were kept, e.g., a yardstick for the linen market, and standardized weights for the grain markets. Instead of the capriciousness of earlier clocks, by this period there emerged an assumption that the bells would ring according to the time as indicated on the sundials, and that clocks used on cloudy days would still be synchronized with those sundials when possible. At this point, regulations that specify the source of a time signal disappear.

St. Mary's Abbey's Bells

In addition to the time pluralism created by the bells controlling the Pavement and Thursday markets, St. Mary's Abbey also played a role in signaling time. Located outside the walls of York, this abbey had its own courts, prisons, and powers separate from those of the city. This abbey was one of the richest and most powerful abbeys in England. Its founding date is unclear,[75] but its wealth can be attributed to royal patronage beginning with King William II. A customary and ordinary for the abbey from the fourteenth century survives[76] and gives a detailed account of the daily cycle of activity and the use of a variety of bells to punctuate that cycle.

The monarchy was not only associated with the abbey, but also with curfews. In fact, the abbey is the only institution recorded as ringing the curfew bell in York during the late medieval period. The sixteenth-century historian of dubious reputation, Polydore Vergil, wrote that William the Conqueror imposed a curfew on England "Item ut ferociam populi ad otium perducetet" (in order to bring the warlike spirit of the people to peace).[77] While there is doubt whether curfew bells were imposed as a means of social control,[78] there is circumstantial evidence that there was a bell with a distinct signal that ended the day that was known as the *curfew bell*, and that it was associated with the Normans. The word *curfew* comes from the Norman French *couvre feu*—to cover fire. In a 1352 document outlining the workday of the stone masons building York Minster, one finds a reference to the curfew bell: during the winter, the masons were to return to work "usque ad pulsacionem companae abbathiae Beatae Mariae quae vocatur le Langebell" (until the ringing of the bell of the abbey of St. Mary's, which is called the Langebell).[79] *Langebell* was another Norman French word used to refer to the curfew bell because of its long peals. Even though the masons and York Minster were politically separate from St. Mary's Abbey, for this particular time signal, the Minster and its masons agreed to use a peal from this abbey. It is possible that this was a legacy of the Norman associations of the abbey, or it might have been for reasons related to the topography of York (as will be discussed shortly), or both.

This abbey was wealthy and powerful enough that, according to its chronicle, it had an early clock: In 1324 "fecit *dominus Stephanus de* Austerwik tunc sacrista fieri orlogium nouum et Campanam" (Master Stephen of Austwick, then sacristan, completed a new clock and bell).[80] The language suggests that this replaced an even earlier clock, although the chronicle only mentions a "campana" (large bell) constructed in 1306, not a clock. It should be noted that Stephen of Austerwik also spent time in Rumburgh, near Norwich—another place in England associated with early English horology.[81]

The customary and ordinary of the abbey refers to the clock only four times, indicating that it was not used to structure the majority of the abbey's daily cycle of activity. The first daily reference to the clock was at the beginning of Prime, which was "Omni tempore anni ad septem de orologio pulsetur" (At all times of the year until seven of the clock is struck).[82] The second was "Ad X de orologio" (until ten of the clock).[83] The third mention referred to the beginning of Compline at "circa octo de le clocke" (around eight of the clock)[84] at which time the community moved from the Great Mass to the daily parliament. The fourth mention of a clock referred to the routine during the week between Palm Sunday and Easter in which "ad tercium ictum horologii vadunt in refectorium ad commune potum" (until the Terce is struck by the clock, they go into the refectory to the common drink).[85]

In addition to the clock, *The Ordinal and Customary of the Abbey of Saint Mary, York* describes a variety of sonic signals. These signals included: clapping hands, a resonating board that would be struck (tabula), a cymbal, a small handbell (*tynte* or *tincte*), a different small bell called the *squilla* (*scilla*) that gave a shrill sound, a collection of small bells (*tintinnabula*), "minor" bells, and the "great" bell. The great bells could be rung in various patterns.

These bells vary in volume, sound, and context of use. This indicates that some signals were more public than others. The tabula was used primarily in the dining hall or in the chapter meeting and struck by one of the leaders of the abbey, such as the prior. For instance, at the end of the chapter meeting, the sacristan struck the tabula.[86] The cymbal seemed to have a similar purpose of signaling nonliturgical gatherings. It indicated the time of meals and the time to wash hands.[87]

The abbey used *squilla*, *tincte*, and *tintinnabula* within the context of rituals. In fact, the *tincte* is identified as a bell for use within the church,[88] and one of its uses was to call the community to prayer.[89] The *squilla* signaled rising to say grace before a meal. Likewise, liturgical practices used tintinnabula. These bells get their name from their jangling, metallic sound—*tintinnum* in Latin.[90] It is unlikely that these small bells could be heard outside of the abbey. These practices were widespread. From the *Regularis Concordia*,[91] a tenth-century rule for English monasteries, it becomes clear that the sound and quality of bells was an important element of identifying which bell was which, and what was the purpose of the bell. This document was part of a Benedictine reform movement in England and was meant to outline the rules of a monastery. The logic behind the bells discussed in the *Regularis* would carry over to communities surrounding monasteries yet reliant on their bells, as is the case with some elements of timekeeping in York. The *Regularis* distinguishes between *tintinnabulum* (little bell) and other bells. For instance, the signal for the beginning of Prime was the use of a tintinnabulum as a signal for the monks only,[92] but for the beginning of the Mass for the lay congregation, a signum, is rung.[93] Signa were large bells that projected sound into the surrounding area. For signaling the monks to begin manual labor, an abbey used a tabula,[94] but used a signum to end work[95]—presumably because its sound would carry into the fields. The *Regularis Concordia* clearly called for specialized bells, and this specialization was reproduced, for it is also found in Lanfranc's eleventh-century *Monastic Constitutions*, such as at daybreak, when after Lauds "the small bell (*parvulum signum*) shall be sounded lightly by the warden of the church,"[96] whereas for summoning people to mass, the large bell (*maius signum*) was used.[97]

What the rules for the abbey demonstrate is a complex sonic code that distinguished sacred from profane activities, and indicated the sort of activity to be carried out. The residents of the abbey would have been familiar with these distinctions and their related sonic qualities. It is not much of a leap to think that such nuanced differentiation of bell signals was part of daily life outside of the abbey's walls.

At St. Mary's Abbey, the use of the clock became included in the mix of sonic signals and was reconciled with the seasonal liturgical variations of the Rule of St. Benedict. Benedict devised a set of liturgical practices tied to particular times of day that would vary with the season. For instance, in chapter 8 of his rule, Benedict described the timing of the Divine Office at night, instructing members of the community to awake at the eighth hour of the night. In contrast, between Easter and the first of November, the time for Vigils ends before daybreak, at which time Lauds should be recited.[98] The seasonal variability of meals is less great, with the main meal consistently occurring at Nones (midafternoon),[99] but there is variation in the timing of labor. From Easter until October, Benedict expected monks to work from Prime until the fourth hour of the day (a bit after midmorning) and then read from the fourth hour until Sext (solar noon). After Sext they rested, and then recited the Nones office after the eighth hour. They would then eat and work until Vespers. Between October and Lent, they would read until Terce, and after the Terce office, work until Nones. During Lent, they read through the end of terce and worked until the end of the tenth hour.[100]

Two things need to be kept in mind when considering how particular religious communities adapted Benedict's Rule. First, there were variations in the rule and the relationship of work to ritual. Second, the canonical hours that structured the daily cycle of activity were approximate.[101] This approximation was not only due to different ways of reckoning such hours, but also to differences in the visible horizon based on terrain and buildings that obstructed one's view. If the horizon is higher than one's location, then the Sun rises later than if it is level with one's field of vision or lower. In effect, in using canonical hours, the landscape contributes to the emergence of time pluralism.

The history of North Yorkshire provides an example of this. Byland Abbey and Rievaulx Abbey were initially only about two miles apart. They were both Cistercian monasteries, which meant that they followed the same rule and the same daily liturgical and work cycles. Moreover, their techniques for reckoning time would have been the same. Still, the timing of their bells was different enough to create confusion and force Byland Abbey to move: "for each convent could hear the bells of the other each hour of the day and night; this was not fitting and could not in any way be allowed to continue."[102] The reason why their bells' timing was so different was because they used canonical hours and had very different eastern and western horizons. The original Byland Abbey was on top of the rim of a valley and had an unobstructed view of both the eastern and western horizons. In contrast, Rievaulx Abbey was at the bottom of the valley. Its eastern horizon was about 27 degrees higher than looking horizontally from the church, and the western horizon was about 14 degrees higher. Since a 15-degree difference is equivalent to an equinoctial hour of daylight, this meant that at the equinoxes the Sun rose almost two hours later at Rievaulx than at Byland, and that the Sun set almost an hour earlier at Rievaulx than at Byland. As a result, since canonical hours divide the duration of daylight into twelve equal segments, and Byland's daylight duration was longer than that of Rievaulx, the hours at Byland would have always been of greater duration than the hours at Rievaulx.

Returning to St. Mary's Abbey canonical hours and its curfew peal, it is possible that institutions in York chose the abbey bells for this role because the location of the abbey is to the west of the city walls and at a slightly lower elevation than the altitude of the city. This would mean that daylight would begin slightly later at the abbey than in much of the city. On the other hand, the abbey is higher than its western horizon with no walls obstructing that horizon. This is even considering the abbey's walls. The base of the walls to the west was about five meters lower than the location of the abbey church, so even accounting for the abbey's walls to the west, which were raised to around fifteen to sixteen feet high on the west side along Marygate in the early fourteenth century,[103] the timing of Vespers and the curfew bell at the abbey would be later than for locations in much of the city of York, particularly if sundown was observed from up in the abbey's bell tower.

Nonresidents of the abbey would also heed its bells for certain purposes, thereby making it part of the soundscape of the city. As already mentioned, the abbey's curfew bell ended the workday of the masons working on York Minster in 1352. The abbey's bells would also signal the Evensong bell of St. Michael's Church, which was within York's city walls, beginning the nighttime period, during which large quantities of food could not be purchased. Its imposing bell tower would allow it, like York Minster, to assert its temporal signals over large swaths of the city over which it had no official control.

York Minster's Bells

York Minster was a powerful landholder and polity within the walls of York. It was the greatest power in Norman York after the king,[104] and after 1276 it often was embroiled in disputes with the civic government of York.

The Minster Close was not only a separate political entity from the city, but it was also not directly under the authority of the Archbishop of York. In fact, much of the archbishop's estate within York was under the authority of the dean and chapter of York Minster, not the archbishop, making York Minster an important landlord within the city. Along with its possessions, it encouraged significant commercial development in the neighborhoods surrounding the Minster Close, but while it had a great deal of influence over these neighborhoods, it did not have direct regulatory authority over them. Yet, one of the features of signaling time by means of bells is that sound travels across boundaries, so there was nothing to prevent the Minster from using its bells to structure the activities and rhythms of those who served it, but who lived and worked outside of the boundaries of the Minster's holdings. Indeed, with their tall belfries and large bells, York Minster and St. Mary's Abbey were equipped to exert influence beyond their political domains.

The construction of York Minster was a major source of employment, and after the plague struck York in the late 1340s, there was a need to deal with the labor shortage resulting from the high death rates. This was a problem in much of Europe, and where employers attempted to keep wages stable, the focus shifted to labor time.[105] For this reason, the description of the workday for masons at York Minster crafted in 1352 focuses on start times, end times, and work breaks,[106] and they represent a more general pattern of wages and workdays varying with the season.[107]

In 1370, another set of regulations structuring the workday of the masons building York Minster Cathedral was put in place. The workday was determined by sufficient daylight for performing work: "als erly als yai may see skilfully by day lyghte for till wyrk; and yai sall stande yar trewly wyrkande atte yair werke all ye day aftyr, als lang als yai may se [*sic*] skillfully for till wyrke, yf yt be alle werkday."[108] On those holy days during which workers labored for half the day, work ceased when "high none smytyn by ye clocke." Other regulations include the length of the meal at Nones. Between Michaelmas (29 September) and the beginning of Lent, the meal break was "ovyr ye space of ye tyme of an houre," and after the meal, the masons could drink in their guildhall for "ye tyme of half a mileway"[109]—a mileway being the time it took to walk one mile. Between Lent and Michaelmas, work began at sunrise and continued until "a mileway byfore ye sone sette."[110] Between Saint Helen's Day (21 May) and the Lammas Fair (31 July to 1 August), the masons could sleep in their hall after the meal at Nones. The text then adds "and all yer tymes and houres sall be reweled bi a bell ordained yare fore,"[111] indicating that the master mason informed workers of the bell(s) regulating their labor.

This document indicates that even after employers used clock time to structure the workday, there were several ways of determining time: the amount of light, the measurement of duration, and the bell identified and controlled by the master mason. The clock only determined high noon, and one should not assume that clocks indicated the duration of a mileway or even an hour. Indeed, the statement that "yer tymes and houres sall be reweled" by a bell identified by the master mason indicates otherwise—that these other time units were not represented by the Minster's clock, but by the bell chosen by the master mason. Time pluralism was necessary for organizing the day's labor.

Occupational Multiplicity

In late medieval York many people participated in the victualing industry.[112] It is possible that this population's behavior motivated civic and guild elites to create regulations that targeted the selling of food. These regulations were of two kinds: ensuring quality and preventing the control and manipulation of supplies.[113] The opening hours of markets reflect the latter goal. Some markets had two opening hours—those for the general buyer, and those for the professional victualler. For instance, regulations in 1393 prevented professional victuallers purchasing food in the Thursday Market until six was struck by St. William's Chapel on Ouse Bridge.[114] This does not reflect when the market opened, but instead was an attempt to keep victuallers from buying up available supplies at the expense of the general population of the city. Even so, there is evidence that professional victuallers defied the regulations.[115]

Such temporal regulation served as a backdrop for another feature of labor; namely, that many artisans engaged in occupational multiplicity, and often these additional occupations involved selling food or running inns.[116] Occupational multiplicity occurs when one source of employment is insufficient to generate the money necessary to meet obligations.[117] There are temporal challenges in pursuing strategies of occupational multiplicity, as my own ethnographic work in Trinidad demonstrates.[118] The primary obstacle is to prevent conflicts between the different productive and social rhythms of the different economic pursuits. In thinking about the time pluralism of York, one of the implications of the complementary powers of guild and civic authorities is how the time pluralism they fostered could be related to the occupational multiplicity in which many artisans engaged.

Due to occupational multiplicity, the multiple time signals did not just indicate the independence of different institutions, but their competition in the regulation of the lives of residents and visitors. Those who engaged in occupational multiplicity often took on second jobs in innkeeping or victualing; this is strikingly similar to what I saw in Trinidad where those engaged in occupational multiplicity tended to run small shops with irregular hours that allowed them to be flexible in dealing with the time demands of others.[119]

In a way, the proximity of the bells of All Saints Pavement, St. William's Chapel, York Minster, and St. Mary's Abbey made juggling the demands of occupational multiplicity easier. Anyone with different occupational demands in the middle of the city would be able to hear these different bells. But when one considers the use of temporal/canonical hours and the topography of York in light of the example of Rievaulx and Byland Abbeys, then it becomes apparent that the choice of bell for regulatory purposes was deliberate.

There is one regulation in particular that at first glance seems puzzling, but when considering the terrain, makes sense. This was the 1393 regulation to restrict the sale of food at night to 18 pennies between the Evensong bell at St. Michael's Church and the morning bell at York Minster.[120]

As already discussed, the main bells that regulated economic activity were All Saints at Pavement and St. William's Chapel. Based on the Historic Towns Atlas series on York,[121] it appears that both buildings likely had restricted views of the eastern and western horizons. In contrast, York Minster had a relatively unimpeded view of the eastern horizon, and was at the highest point in the city, and St. Mary's Abbey had an unimpeded view of the western horizon. If the day was determined by somebody in the bell towers of the Minster's and the abbey's churches, then the view of the respective horizons was even more unimpeded. This meant that the morning bell of the Minster tolled before the morning bell of All Saints or St. William's Chapel. St. Michael's Church was an extension of the abbey, so its bell was rung in response to the abbey bell. Since the Evensong bell of the abbey would have been one of the last Evensong bells to ring in the city, St. Michael's Evensong bell would have been particularly late in the day, as well—signaling the beginning of nightfall after both All Saints in Pavement and St. William's Chapel.

This leads me to surmise that the choice of St. Michael's and York Minster for the regulation restricting the sale of food at night took advantage of time pluralism to foster occupational multiplicity. The Minster bell would have provided a window in the morning during which one could buy and sell quantities of food before the markets opened according to bells rung at St. William's Chapel or All Saints. Likewise, St. Michael's Evensong bell would have struck after a window of time that followed the ending of the workday according to All Saints or St. William's Chapel.

Cacophony

It is not surprising, then, that the regulations that emerge during the fourteenth through sixteenth centuries in York reflect a diversity of interests demonstrating a time pluralism that must have been bewildering to visitors. It is not enough to document all the different sources of time, however. One must also try to understand how people in York discerned between these different sources. What were the codes and contexts that allowed people to avoid too much confusion in determining the time relevant to their activities?

The specification of a particular bell is quite unlike the current practice, which relies on the idea of a single timescale represented on synchronized clocks to coordinate and regulate activity. The need to choose a bell, and consequently a particular institution's reckoning of time, indicates deference toward one institution and source of power among many within the city. In a city such as York, where there was a mixture of guilds, priories, city administration, a powerful abbey, a separate administration of York Minster's precinct, and added to this, the power of individual employers to pick and choose their source of time, it would seem that the potential for confusion would be great. Such confusion is documented in other cities in Europe.[122] Yet, for York there is no documentary indication of confusion over time, only clear indications of which institution's time governed particular activities. In effect, it seems that what mattered was knowing who had power over one's activities, and from that followed the source of time to which one adhered.

The economic and political complexity of the city contributed to the sonic complexity of time pluralism. Throughout Europe, the variable performance of public clocks also contributed. As late as 1657, one finds discussions of public clocks within a city disagreeing. In a dialogue written by Anthony van Torre for the instruction of Jesuits, the preceptor asks, "Ex quo horologio te dirigis?" (From what clock do you get your time?). His student responds by naming several clocks. The preceptor's response is then "Fere quotidie dissonant horologia: praeveniendus est error diligentia." (Almost every day the clocks disagree: anticipate this error by means of diligence).[123]

The late medieval period through the early modern era were not times when synchronization of clocks was common, and when attempted, it met with poor results. In Paris, King Charles V of France decreed that all the bells of the city would strike in unison with the bells driven by the clock in the royal palace. In 1370, he had a clock constructed in the Palais de Justice by Henry de Vic and described in detail by Ferdinand Berthoud.[124] The clock was meant to encourage "the orderly running of the bureaucracy and orderly life of the citizens,"[125] but it also substituted equinoctial secular (in the sense of worldly) hours for the canonical hours.[126] While there is doubt that there ever was a formal decree to synchronize Paris's clocks to this particular clock,[127] the stature given this clock by the king may have given rise to the expression "C'est l'horloge du Palais; elle va comme ça lui plait" (This is the clock of the palace; it goes as it pleases). While the king's clock maybe went as it pleased, it differed in a crucial respect from all other clocks that kept their own, unique time: it was the king's clock. Still, the expression and surviving record do not indicate that Charles V was able to successfully synchronize and regulate life in Paris. Paris's clocks defied any attempt to impose an idea of a single timekeeper.

Based on the Palais de Justice clock, it is easy to attribute the lack of synchronization solely to the lack of accuracy of clocks, but I think this misconstrues the associations of power with bells and clocks and an anachronistic projection of the ideas of accuracy and synchronization back in time. The desire for synchronization is modern and related to the idea of the ideal clock. Moreover, synchronization disempowers local authorities who previously could control people by means of controlling clocks. E. P. Thompson mentions this in his article on work discipline and industrial capitalism: employers discouraged workers from carrying watches because that would give the workers a means of challenging the employer's clock and power to manipulate the clock.[128] Thompson writes about the eighteenth century, but employers' control over time by controlling clocks was not an eighteenth-century innovation—it was a long-standing practice. In examining material from Antwerp, Bruno Blondé and Gerrit Verhoeven describe the emergence of clock-driven time awareness as "'slow burn' evolution."[129] The innovation of the eighteenth century was reliable timekeeping that made challenges to poorly performing clocks possible and that exposed the power latent in controlling a clock.

With this in mind, in a city in which there were competing political interests, lack of synchronization was not a reflection of poor time-keeping practices, but a symptom of political competition. Indeed, in the history of York, the advent of reliable timekeeping diminished the importance of individual clocks. By the late seventeenth century, the markets had market crosses with sundials, and within the subsequent hundred years, those sundials were replaced with clocks. Starting with this period, regulations no longer mention a specific time source, only a time.

In thinking about the soundscape of late medieval York, one has to keep in mind that due to differences of visible horizons, and the possible differences in judgment over what defined the moment of dawn and dusk, the four major political powers would have reckoned canonical hours differently, thereby affecting the ringing of their respective bells. There was not, nor could there have been, synchronization between these bells without one following the other, but these political entities were too protective of their rights to have deferred in such a way. Using one's imagination of what it was like, one would have probably heard the morning bell at York Minster first, followed by the morning bell at All Saints Pavement, followed by the morning bell at the abbey, and lastly the morning bell at Ouse Bridge. All these bells would agree at solar noon, but then they would diverge again. The key hour of Nones would ring first at Ouse Bridge, second at All Saints, and then it is likely that the Minster bell would have proceeded the abbey bell for this hour because it would have had a more built-up western horizon—although that is difficult to determine with certainty without accurate information on the bell towers in the fourteenth century. The Vespers bells would ring first at Ouse Bridge, second at Pavement, probably third at the Minster, and lastly at the abbey, and the last bell of the day, the curfew bell, would ring at the abbey. Each of the city's powers had its own reckoning of canonical hours such that the time referred to in any regulation also included a specification of the time source.

Timescale pluralism in York, then, was not just a matter of power, but of a privileging of locations, institutions, and techniques for reckoning time. This was in the interest of these powers and places, not necessarily in the interest of all tasks done within the city. The common argument that before clocks time was solely task oriented is thus false. The non-clock, multiple ways of reckoning time reflected power just as much as clock time. To those who must heed its signal, a bell rung by hand is no different than a bell rung by a clock. In a way, the non-clock time signals were more direct, less mediated forms of power than later clock time. One knew the power behind the timescale, whereas with modern clocks, one only knows the power that defines rules and regulations in relationship to the timescale. There are few who can directly define and control a timescale today.

Notes

1. Glissant, *Caribbean Discourse*, 106.
2. Anonymous, *Here endith*, chapter 18.
3. Marc Bloch, *Feudal Society*, 73–74.
4. Whitrow, *Time in History*, 17.
5. Rossi, *Philosophy, Technology and the Arts*, 36.
6. Crosby, *Measure of Reality*, 27.
7. Febvre, *Problem of Unbelief*, 393–95.
8. Le Goff, *Time, Work, and Culture*, 175.
9. Birth, *Objects of Time*, 55–66, and "Capital Flows"; Gimpel, *The Medieval Machine*, 108.

10. Bradbury and Collette, "Changing Time"; Dohrn-van Rossum, *History of the Hour.*
11. See Corbin, *Village Bells*, 117–18.
12. Atkinson, *Noisy Renaissance*, 147.
13. Dohrn-van Rossum, "Uhren und Uhrzeiten," 86–87.
14. Dohrn-van Rossum, "Uhren und Uhrzeiten," 86.
15. Glennie and Thrift, *Shaping the Day*, 37.
16. Burke, "Cultural History of Time," 620.
17. Rooney, *About Time*, 66.
18. Wilkin, "Time Constraints."
19. Davis, *Medieval Market Morality*, 182–83.
20. Humphrey, "Time and Urban Culture," 107.
21. Brand, "Lawyers' Time in England," 90.
22. Hall, "Earliest Record"; Rogers and Hall, "Striking Discovery."
23. Colvin, *Building Accounts*, 433.
24. Vilhjalmsson, "Time and Travel."
25. Scott and Cowham, *Time Reckoning*, 10–12.
26. See Frake. "Cognitive Maps."
27. See Jones, "Latin Ecclesiastical Calendar"; Hughes, "Implicit Carolingian Tidal Data."
28. Adam, *Timescapes of Modernity*, 11.
29. Garrioch, "Sounds of the City," 6.
30. Atkinson, *Noisy Renaissance*, 11.
31. Garrioch, "Sounds of the City," 60.
32. Dohrn-van Rossum, *History of the Hour*, 209.
33. Potter, *The Bells and Bellringers*, 47.
34. Murphy, "Receiving Royals."
35. Widdington, *Analecta Eboracensia*, xv.
36. Richards, *Bedern Foundry.*
37. Benson, *York Bellfounders.*
38. Rees Jones and Palliser, "York, 1271-1536," 41.
39. Palliser, "York's Civic Liberties," 98.
40. Rees Jones, "York's Civic Administration," 126.
41. Arnold and Goodson, "Resounding Community," 127.
42. Miller, "Medieval York," 92.
43. Richardson, "Guilds, Laws, and Markets," 16; Basing, *Trades and Crafts*, 63–64; Miller and Hatcher, *Medieval England*, 57.
44. Le Goff, *Time, Work, and Culture,* 44; *Medieval Civilization*, 182.
45. Bartholomaeus Anglicus, *De genuinis rerum coelestium*, book 9, chapter 23.
46. Bartholomaeus Anglicus, *De genuinis rerum coelestium*, book 9, chapter 24.
47. Sellers, *The York Memorandum Book*, 93.
48. quoted in Wade and Brožek, *Purkinje's Vision*, 109–10.
49. Miller, "Medieval York," 98.
50. Thrupp, *Merchant Class*, 93.
51. Swanson, "Illusion of Economic Structure," 29.
52. Swanson, "Illusion of Economic Structure," 31.
53. Page, *Inventories of Church Goods,* 90.
54. Rees Jones, *York*, 43.
55. Palliser, *Medieval York*, 117.
56. Sellers, *York Memorandum Book*, 49.
57. Sellers, *York Memorandum Book*, 172.

58. Raine, *York Civic Records, volume III*, 67.
59. Raine, *York Civic Records volume III*, 67.
60. Allison, "Markets and Fairs," 486.
61. Rees Jones, *York*, 263–64.
62. Swanson, *Medieval Artisans*, 11; Rees-Jones, *York*, 265.
63. Drake, *Eboracum*, 242.
64. Rees Jones, "York's Civic Administration," 126.
65. Orme, "Church and Chapel," 79.
66. Wilson and Mee, *"Fairest Arch in England,"* 39–48.
67. Vitruvius, *Ten Books on Architecture.*
68. Howard and Moretti, *Sound and Space*, 6.
69. Kircher, *Phonurgia Nova*, 34.
70. Boethius, *Fundamentals of Music*, 1–14.
71. Rimple, "Enduring Legacy."
72. Adelard of Bath, *Conversations.*
73. Raine, *York Civic Records, volume V*, 102.
74. Raine, *York Civic Records, volume V*, 148.
75. Knight, *History of the City*, 135–36.
76. Tolhurst, *Ordinal and Customary.*
77. Polydore Vergil, *Urbinatis Anglicae Historiae*, 151.
78. Cuming, *Proceedings*; Barber, "Norman Conquest," 19–20.
79. York Minster, *Fabric Rolls*, 172.
80. Craster and Thornton, *Chronicle of St. Mary's Abbey*, 73 emphasis in original.
81. North, *God's Clockmaker*, 153.
82. Tolhurst, *Ordinal and Customary*, 49.
83. Tolhurst, *Ordinal and Customary*, 87.
84. Tolhurst, *Ordinal and Customary*, 158.
85. Tolhurst, *Ordinal and Customary*, 281.
86. Tolhurst, *Ordinal and Customary*, 285.
87. Tolhurst, *Ordinal and Customary*, 142.
88. Tolhurst, *Ordinal and Customary*, 68.
89. Tolhurst, *Ordinal and Customary*, 107.
90. Rocca, *De Campanis Commentarius*, 16.
91. Aethelwold, *Regularis Concordia.*
92. Aethelwold, *Regularis Concordia*, 14–15.
93. Aethelwold, *Regularis Concordia*, 19–20.
94. Aethelwold, *Regularis Concordia*, 20.
95. Aethelwold, *Regularis Concordia*, 21.
96. Lanfranc, *Monastic Constitutions*, 3.
97. Lanfranc, *Monastic Constitutions*, 4.
98. Benedict, *Rule*, chapters 20–21.
99. Benedict, *Rule*, chapter 41.
100. Benedict, *Rule*, chapter 48.
101. Rothwell, "Hours of the Day."
102. Burton, *Abbeys of Byland and Jervaulx*, 13.
103. Royal Commission on Historical Monuments, *Inventory*, 166.
104. Palliser, *Medieval York*, 148.
105. Dohrn-van Rossum, "Uhren, Glocken und Zeitorganisation," 76.
106. See Birth, *Objects of Time*, 56–65.
107. Epstein, *Wage Labor*, 113; Birth, "Capital Flows."

108. York Minster, *Fabric Rolls*, 181.
109. York Minster, *Fabric Rolls*, 182.
110. York Minster, *Fabric Rolls*, 182.
111. York Minster, *Fabric Rolls*, 182.
112. Swanson, *Medieval Artisans*, 9.
113. Swanson, *Medieval Artisans*, 9.
114. Drake, *Eboracum*, 242.
115. Swanson, *Medieval Artisans*, 10–11.
116. Swanson, "Illusion of Economic Structure," 34–36, and *Medieval Artisans*, 6.
117. Comitas, "Occupational Multiplicity."
118. Birth, *Trinidad Time.*
119. Birth, *Trinidad Time,* and *Objects of Time.*
120. Drake, *Eboracum*, 242.
121. Addyman, *York.*
122. Marpergers, *Horologiographia*, 33.
123. van Torre, *Dialogi Familiares*, 120.
124. Berthoud, *Histoire de la mesure*, 66–70.
125. Dohrn-van Rossum, *History of the Hour*, 156.
126. Attali, *Histoires du temps*, 129.
127. Dohrn-van Rossum, *History of the Hour*, 217–19.
128. Thompson, "Industrial Capitalism," 86.
129. Blondé and Verhoeven, "Against the Clock," 236.

Chapter 3

Cosmopolitan Times

Time Pluralism in Habsburg Horology

> We point out to the reader that there are three kinds of time-reckoning: it operates either according to nature, or according to custom, or according to authority.[1]
>
> —Bede, *Reckoning of Time*

If anyone could impose a single timescale, it would be an emperor. Yet, despite being a collector of clocks, the Holy Roman emperor Charles V did not mandate a timescale. Instead, contrary to our current practices, which emphasis uniformity, in the Holy Roman Empire many public clocks indicated multiple timescales, reflecting the diversity of timekeeping and the imperial attempt to embrace cosmopolitanism rather than uniformity.[2] In other words, narratives of the emergence of the current timescale obscure a fundamental ideological shift from emphasizing cosmopolitan timekeeping, and hence time pluralism, to emphasizing uniformity in timekeeping.

Charles V was a pivotal figure in Europe. He was the last Holy Roman Emperor who envisioned a unified Christian Europe,[3] but he was also the emperor who ruled as Europe split apart during the Reformation. Due to Spanish incursions into the New World, Charles oversaw European colonial expansion in unprecedented ways. He collected art and curiosities, but the only objects in Charles's collections that really interested him were the clocks.[4]

His obsession with timekeeping complemented his colonizing and imperializing policies dependent on cartography, navigation, and astronomy.[5] His sponsoring voyages of exploration were as much about mapping and claiming parts of the world as about discovering them. Such mapping depended on timekeeping—particularly on the use of ephemerides to know celestial arrangements for a known location and time and then compare them to the explorer's location. This allowed the rough calculation of longitude on land.

When Charles V abdicated the throne in 1556, he retired to the Monastery of Yuste with his clockmaker Torriano, also called Juanelo. Charles's rooms contained many clocks, some of them with complicated displays that included astrological information and different hour systems. With these were other astronomical instruments, and maps of different parts of the world, including the Indies.[6] Charles's days in retirement began with a visit from Torriano the clockmaker, and together they would spend much of the day trying to get his many clocks to synchronize.[7] Charles's son, Philip II, expanded the clock collection as part of his extensive collection of fine art,[8] books, and objects of natural history that he strove to make the greatest in the world.[9] Such collections were created in two ways: patronage of artisans and gifts received from other nobles or artisans seeking support.

The Emphases of Habsburg Horology

Patronage of clock making influenced horological design and function. During the early modern period, Habsburg ideas of accuracy subtly differed from those of today. This was because of an emphasis on determining the accuracy of a particular moment in time for use in astrology versus the modern emphasis on determining the accuracy of a unit of measured duration for use in regulations.

Habsburg gift clocks emphasized the representation of celestial cycles. This was indicated in several ways on the clocks. Most explicitly, it was the reason for including a dial modeled on one side of an astrolabe on many clocks. This dial indicated the movement of the fixed stars, Sun, and Moon for a particular latitude. When one related the astrolabe dial to dials indicating the dominant planets for each day of the week and each hour of the day, the clock became a robust astrological tool. While the Church condemned astrology that predicted the course of human events, it permitted other forms of astrology, such as meteorological and natal astrology.[10] During this period, scholarly ideas about how the cosmos functioned included astrological principles.[11] In fact, some viewed astrology as a science that unified all other areas of inquiry.[12]

Habsburg emphases on astrology and cosmopolitan approaches to different hour systems reflected Habsburg politics. Historians describe the Habsburg Empire as a conglomerate state and as a composite state. Conglomerate states consist of different territories with different levels of autonomy and varying relations to the state apparatus.[13] Composite states involve multiple polities under a single rule.[14] The Habsburg Empire involved both features with free cities, and independent states all being ruled by the Holy Roman Emperor. As emperor, Charles had to work with the Imperial Diet, a body consisting of the various territorial princes.[15] He also had many vassals who tended to act independently.[16] There were corners of the empire over which Charles had little control.[17] Regionalism, rather than centralization, characterized the empire,[18] and political authority in the empire "hover[ed] awkwardly between the imperial and the provincial,"[19] making the power of the monarch dependent on the cooperation of local nobility and officials.[20] In this situation, Charles relied on family ties and patronage to manage imperial challenges. The historian H. G. Koenigsberger described the political structure as thus: "The king no longer stood at the apex of a feudal pyramid but is best visualized as sitting, spider-like, at the centre of a kingdom-wide network of patron-client relations."[21] The multiple regional and provincial timescales found within Charles's empire were a sign and symptom of such political decentralization and local autonomy.

The Habsburgs were distributed across Europe, and they spoke different languages and participated in different cultural traditions. As a result, the empire's elite developed a cosmopolitan quality.[22] Such diversity was a feature of imperial rule established by the time of Charles's father, Maximilian,[23] and it provided the foundation for Charles's court to be viewed as "the most cosmopolitan in the world."[24] In such a situation, courting cosmopolitanism was a strategy for keeping divergent polities together. Moreover, such a loose conglomeration of polities could only be kept together with strong systems of patronage.[25] The multiplicity and cosmopolitanism of Habsburg timescales reflected this.

Many clock dials were made in this empire to display different hour systems, and one design was common. The outer ring of the dial contained the small hours—two counts of twelve hours beginning at midnight. The next ring showed Italian hours, which began when the Angelus was rung after sundown. This ring linked to a calendar gear so that it adjusted for the seasonal changes in sundown throughout the year. The next ring was also linked to the calendar gear and represented the great or Nuremberg hours. Since this system had a different count of hours for daytime and nighttime, it posed a different mechanical challenge that was met with a shutter system that would expand and contract based on the day of the year. Finally, the innermost ring displayed the Babylonian hours, which began at sunrise. Thus, such clocks displayed four different hour systems in relationship to one another, thereby allowing for the easy conversion between them (see figure 3.1).

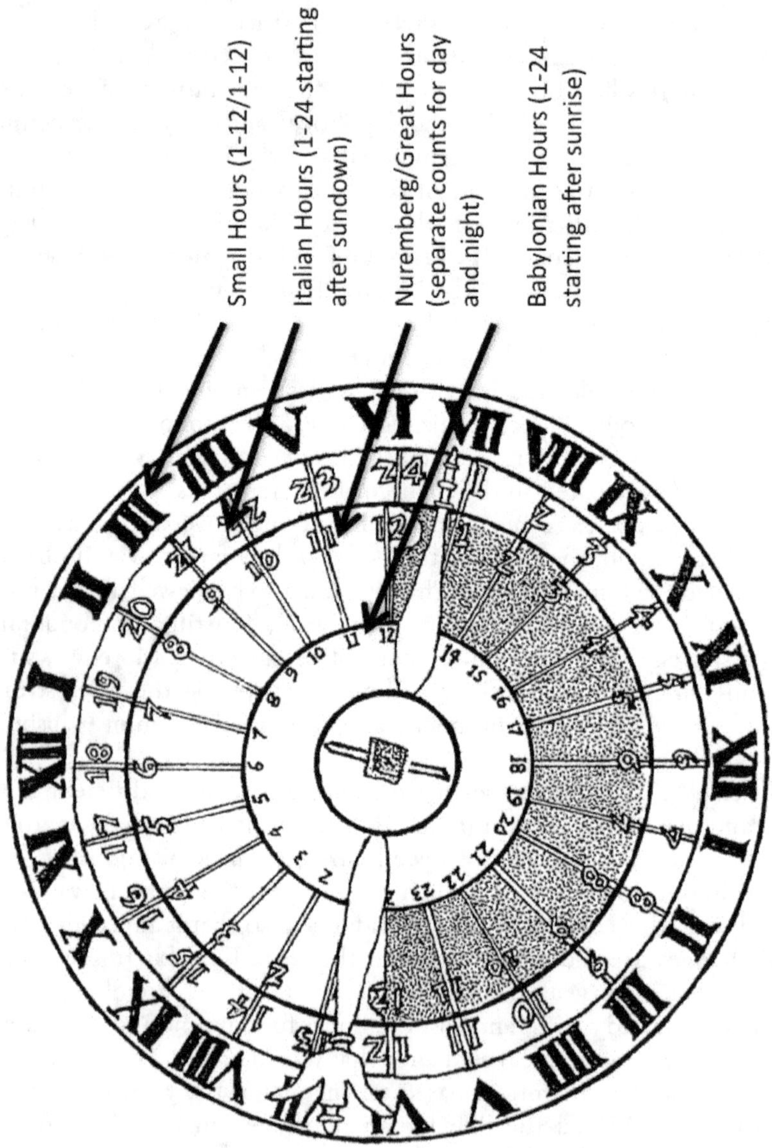

Figure 3.1. Clock dial on the Weber masterpiece clock in The Frick Collection. © Kevin K. Birth.

The diversity of timescales displayed by these clocks should be interpreted in this light. The display of the different hours would not only demonstrate the cosmopolitanism of the owner, but also acknowledge the diversity the owner's kin, friends, and guests.

Clocks as Gifts; Gifts as Reproducing Knowledge

The era of Charles V was the period of the emergence of *Kunstkammers* and *Wunderkammers*. These were collections of art, natural history specimens, automata, and scientific instruments. The historian of science, Paula Findlen, writes of these collections: "Through the possession of objects, one physically acquired knowledge, and through their display, one symbolically acquired the honor and reputation that all men of learning cultivated."[26] These collections not only involved the amassing of objects and information, but they were also nodes in social networks involved in the creation and distribution of knowledge.[27] They brought together elite patrons, humanists, scientists, and artisans, which resulted in personal ties and friendships that were important in the improvement of mathematics and astronomy.[28] Yet, these networks created knowledge under the gaze of political elites. In the case of technologies and ideas of timekeeping, they were sites where mechanical expertise, astronomical observation, philosophical imagination, and financial support intersected. The scientific instruments in collections, including clocks, were not just matters of entertainment or curiosity, but they were also colonial and political tools for commerce and regulation,[29] as well as for impressing the public.[30]

Samuel Quiccheberg's *Inscriptiones vel tituli theatric amplissimi* guided many collectors. He stated that such collections should include clocks.[31] Many collectors also emulated the example of the Duke de Berry, whose early and famous collection included clocks.[32] Since several of the common clock designs had multiple hour systems, the practice of collecting these clocks distributed these hour systems beyond their place of origin.

Often, the contents of collections became widely known even as many people had no access to them. Some collectors, such as the Holy Roman Emperor Rudolf II, carefully restricted access and used such limited access to flatter high-ranking visitors, including ambassadors.[33] In addition, to gain an audience with Rudolf II, visitors would give him a gift for his Kunstkammer.[34] This entangled diplomacy with collecting, science, and technology. In contrast with Rudolf, other collectors viewed large numbers of people visiting their collection as a sign of success,[35] and as a means of expanding their social network:[36] Findlen writes, "If admission to one's home or museum signaled the conferral of friendship, then this was the tie that bound together collectors and visitors all over Europe. Such exchanges reaffirmed the 'reality' of the republic of letters by defining the social parameters of that world."[37] Collections either were a means of demonstrating prestige through controlling access or of building prestige through encouraging visits. Both practices depended on cultivating the reputation of the collection.

In such collections, science and technology became vehicles for asserting a patron's standing, and at the same time, ways of recognizing reputations of the givers and recipients of clocks. Gift clocks mediated relationships and distilled dialogues about time and technology. These were complex dialogues. Such exchanges were attempts at a carefully orchestrated exchange involving tensions between cosmopolitanism, appropriation, displays of difference, attempts at standardization, and distribution of knowledge. Moreover, structured by the rubric of exchange in which the giver and receiver were both known, they unfolded in a fashion far different from the current system in which clocks mediate between users and a set of institutions that calculate and distribute the time.

Patronage, Exchange, and Science

The example of the clock made by Baldewin for Landgrave Wilhelm IV shows the entanglement of diplomacy, technology, science, and patronage. The story of this clock begins with the book *Astronomicum Caesaraeum* by Peter Apian.[38] Apian dedicated this book to Charles V, the aforementioned emperor known for his affinity for clocks and his patronage of horologically related science. It is no surprise that books about astronomy and timekeeping were dedicated to Emperor Charles V. Machiavelli articulated the logic of book dedications in his own dedication of *The Prince* to Lorenzo de' Medici, when he wrote, "In most instances it is customary for those who desire to win the favour of a prince to present themselves to him along with those things which they value most or which they feel will most please him."[39] A dedication to a book was viewed as a type of gift,[40] and such gifts were not merely an offering of an object, but also of knowledge. Artisans and scientists offered knowledge and objects embodying that knowledge in hopes of obtaining material support in return.

Apian's book described the design and use of astronomical instruments, including ways of marking time. It included color images of discs representing important celestial cycles, such as those of the Sun, the Moon, and the planets. In a way, because of this attempt to explain related but different cycles, it embodied the polyrhythms that underlay Renaissance and early modern timescale pluralism.

This book caught the attention of Landgrave Wilhelm IV of Hesse-Kassel. Wilhelm IV was a skilled astronomer with a penchant for new astronomical instruments. His collections included 442 scientific instruments, among them clocks. These were not merely for display, but for use.[41] The work of the astronomers Regiomontanus and Bernhard Walther of a century before inspired Wilhelm.[42] Like Regiomontanus and Walther, the inaccuracies and inconsistencies in existing astronomical tables frustrated Wilhelm.[43] His awareness of Regiomontanus and Walther likely led him to hire Andreas Schöner as his astronomer. Andreas Schöner was from Nuremberg, the adopted home of Regiomontanus and Walther, and was the son of Johannes Schöner, who had acquired Regiomontanus's and Walther's astronomical observations and published them in 1544.[44]

The images in Apian's book led Wilhelm IV to conceive a clock with dials representing the information in Apian's images. Wilhelm worked with Andreas Schöner to design this clock. The notes and tables associated with these efforts are bound with a copy of Apian's book, which is housed in the library of the University of Kassel and made available online,[45] appearing as pages preceding and following Apian's text.[46] Wilhelm IV then commissioned Eberhart Baldewin to make the clock. In a case of how a powerful patron can shape research questions and the design of scientific instruments,[47] Wilhelm's desires shaped the relationship between Schöner's astronomical work and Baldewin's technical and mechanical work.

Baldewin was not a clockmaker, but a skilled mechanic. He served as the *Lichtkammer* of Wilhelm's brother, Ludwig. A Lichtkammer maintained the heating and lighting for the landgrave's residence.[48] Period treatises on machinery suggest that the sort of mechanical expertise that Baldewin had involved a thorough knowledge of gears and gearing,[49] which equipped him for the clock-making task he was given. Baldewin's skill in designing and making machines must have caught Wilhelm's attention. Baldewin's talents also included sketching and communicating his designs to Wilhelm, which was critical in working with Wilhelm to bring the clock to fruition.[50]

The clock that Baldewin constructed represented astronomical cycles from a Ptolemaic, geocentric perspective, and included the cycles of the Sun, the Moon, and the planets—each with its own dial. The clock also included an automated astrolabe designed for 38 2/3° south of the pole, which is 52 1/3° north latitude (the latitude of Kassel) to represent the Sun's cycles and the movement of the major fixed stars.[51] Not only did it assist in reckoning time, but it enhanced Wilhelm's fame with other princes.[52] According to Karsten Gaulke, curator of the astronomy collections at Kassel, Wilhelm's scientific reputation "was a new way to gain influence and power in the diversified political landscape of the Holy Roman Empire."[53]

Wilhelm IV must have been quite satisfied with his clock, for a few years later he commissioned Baldewin to make a similar clock to be given to

Augustus I, Elector of Saxony.[54] Augustus was another patron of astronomical and horological science, having accumulated a collection of instruments and clocks that became the foundation for the Mathematisch-Physikalischer Salon in Dresden. In fact, it seems that many items in Augustus's collection depended on gifts from Wilhelm IV.[55] But Augustus did more than collect. His patronage developed a scientific instrument-making industry in Saxony.[56] Between the gifts he received, and the clocks made in Saxony, by 1587 there were thirty-five clocks in his Kunstkammer and one hundred other astronomical/time-measuring instruments.[57] In contrast to the present metrological emphasis on uniformity, Augustus and Wilhelm were de facto patrons of developing multiple ways of representing time.

In the dedication of his book on sundials and astrolabes, Andreas Schöner praised such princes who encouraged the design of scientific instruments.[58] The involvement of princes in the support of scientific-instrument design and manufacturing was a trend in the sixteenth-century Habsburg domains.[59] This support was crucial in the development and distribution of time-keeping ideas, but these ideas reflect the time pluralism of Wilhelm's domains with the emphasis on relating different hour systems and cycles.

Distribution of Science and Technology

Scientific instruments, including clocks, reflected the interests of the patrons who sponsored their production,[60] and could shape the issues that scientists researched.[61] Often, we think of inventions as fulfilling existing needs, and such a case has been made for the invention of the mechanical clock,[62] but once invented, technologies can become resistant to change and thereby come to shape needs rather than meet them—as Lynn White put it, "Human life no longer adapts the mechanism to its needs; mankind is in some measure shaped by a machine which it adores."[63] The relationship of patron and client was one of the clients' "self-fashioning" themselves to fit the social context and desires of the patron.[64] In turn, the products of these client-scientists would be used by their patrons as gifts that not only acknowledged relationships, but de facto circulated ideas, as well. Now, clocks hide relationships so that few are aware of any relationship to the metrologists who manage UTC.

Some histories of science and technology focus on how technologies follow one another, while neglecting how ideas became adopted and disseminated.[65] If one focuses on the distribution of technology rather than just its development, then with timekeeping in the early modern period one could argue that it was the science sponsored by elites within the Habsburg domains that encouraged the widespread distribution of scientific instruments related to timekeeping and surveying, and that this fostered the first wave of European colonial expansion and land claims.[66] In these technologies and techniques, timekeeping and mapping were intertwined. Claims of discovery led to claims to territories, but such claims were not sufficient without means to validate the claims. Some of these means involved cartography and surveying: documenting the location of land in terms of latitude, which could be ascertained with time-reckoning instruments like the astrolabe.[67] Maps became a "visual language communicating proprietorial or territorial rights in both practical and symbolic senses."[68] Maps represented frontiers.[69] While longitude is difficult to determine at sea, the first great colonial division of "discovered" territories—the Treaty of Tordesillas signed in 1494—referred to it. This treaty defined a meridian 370 leagues west of the Cape Verde Islands as the division between Spanish and Portuguese claims. Moreover, the earliest known map of geopolitical boundaries, produced in 1502, displayed this division of the world.[70] The treaty of Ardres, between England and France, used a map in the negotiating process.[71] Maps developed a focus on defining boundaries, thereby becoming a "language of property."[72] Such precise managing and division of space was made possible by new instruments, surveying techniques, and developments in cartography:[73] "when Europeans were mapping America for the first time, a new paradigm in cartography was simultaneously being diffused ...were it not for the new maps, America would hardly have entered the European consciousness and could not have been colonized in the way it was."[74] Moreover, during this period, governments began to use maps as administrative tools.[75]

Time-keeping ideas played an essential part in this process by integrating astronomical timekeeping that was sensitive to longitude and latitude with mapmaking and navigation.[76] By the early fifteenth century, mapmakers used latitude and longitude tables.[77] Determining longitude on land involved using celestial observation to ascertain the time difference between one's local meridian and the longitude for which tables were made. Determining latitude depended on the altitude of the Sun at noon, or the celestial pole at night, and was essential to the proper functioning of a sundial. The process of standardization and homogenizing of space and time in cartography granted legitimacy to the maps produced.[78] This involved the development of conventions and simplification that permitted their circulation among experts without questioning their representational value.[79]

The fusion of mapmaking and astronomy was accelerated by the translation of Ptolemy's works into Latin in the mid-fifteenth century—an effort associated with Johannes Regiomontanus and his mentors Georg von Peuerbach and Cardinal Bessarion. Bessarion, a native speaker of Greek, reputedly learned Latin so that he could translate Ptolemy's works into that language.[80] Bessarion encouraged Regiomontanus to learn Greek in order to finish Peuerbach's abridgment of Ptolemy's work on astronomy, the *Almagest*.[81] Ptolemy had developed the system of defining locations on a world map in terms of longitude and latitude.[82] He had discussed latitude in his *Almagest*,[83] and both latitude and longitude in his *Geography*,[84] which provided longitudes for the world known to Europeans, although these longitudes were incorrect.[85] Anticipating the eventual determination of longitude by means of using a chronometer, Ptolemy had also related the equinoctial hours to intervals of 15 degrees of longitude.[86] Regiomontanus was instrumental in disseminating Ptolemy's *Almagest* and had written that he also wanted to do the same for his *Geography* as well as publish maps.[87] Consequently, by the end of the fifteenth century, while Columbus was not the first European in the Americas, nevertheless he used navigation and astronomy to define certain territorial claims in ways that previous European sailors had not.

Cosmopolitanism and time pluralism encouraged patronage of science and horology that then enabled technology and techniques crucial to the administration of expanding empires. This approach to governing large, diverse polities that included colonies emphasized timescale pluralism. This emphasis was a demonstration of the power to rule despite local differences. It is a very different logic from modernity, in which power is closely tied to standardization and regulation.

The Case of Rudolf II's Court

As pointed out earlier, dedications of books were a form of gift, and the gift of a book was meant to give knowledge in hopes of support. Time pluralism, timekeeping for scientific observations, and the politics of empire became intertwined. Such was the case with Rudolf II's sponsorship of Tycho Brahe and Jost Bürgi.

When Tycho Brahe had lost his beloved observatory of Uraniborg due to changing court politics in Denmark, he sought a new patron. There were several possible candidates, including Landgrave Wilhelm IV whom Brahe had already visited and advised. Indeed, Wilhelm had astronomical instruments constructed based on Brahe's recommendations and designs. But Brahe aimed higher. He knew of Rudolf II's interest in astronomy and observational instruments, so Brahe dedicated his book on this topic to Rudolf as an attempt to ingratiate himself to the monarch.[88] Brahe's effort worked, and Rudolf invited Brahe to become the court astronomer and offered to set him up with his own observatory outside of Prague.

Brahe's dedication reviewed the history of astronomical observation and the instruments, with such telling comments as noting that science advanced "ubi Imperatores & Reges sumptus talibus Organis conficiendis liberaliter suppeditarunt" (where emperors and kings freely supplied completed instruments when needed).[89] Brahe also discussed the instruments he employed with which he would have known that Rudolf II had some familiarity—such as armillary spheres and quadrants. Brahe concluded his dedication by saying that his book would discuss instruments and techniques he invented,[90] and by praising Rudolf II's knowledge of astronomy and with the hope that the emperor would look upon Brahe's efforts favorably.[91]

To achieve his goals, Brahe relied on a variety of instruments and constantly honed his time-keeping techniques. He measured right ascension in terms of time, not degrees. The level of time-keeping precision to which he aspired in his records was the second. In effect, his effort at timekeeping was to map the heavens, and for an astrologically oriented monarch like Rudolf II, mapping the movement of the heavens was a step toward creating clocks that could chart such movements. Before Brahe's arrival in Prague, Rudolf II had already sponsored the construction of celestial globe clocks that modeled the heavens. Such celestial globe clocks were different from the clocks Baldewin had made for Landgrave Wilhelm IV. Instead of displaying a different dial for each planet, the celestial globe clocks worked like armillary spheres in visually depicting the movement of objects in the sky rather than representing their movement in terms of the duration of their cycles.

In 1604, a couple of years after Brahe's death, Rudolf II employed Jost Bürgi, an innovative Swiss clockmaker who previously had served Landgrave Wilhelm IV of Hesse, who had already demonstrated through his patronage of Baldewin the tendency to turn horological developments in his court into gifts of clocks to other nobles. Under Wilhelm's patronage, Bürgi developed a means of determining the longitude of stars by using a clock instead of taking angular measurements.[92] Consistent with Rudolf II's tastes, Bürgi is most famous for his outstanding celestial globe clocks. Bürgi made other clocks, improving on escapements and the length of time a clock could run. He also was the first to make a clock that could measure seconds.[93] To accomplish the latter, he developed the first remontoir—a self-winding mechanism. While his innovations improved the performance of clocks, "he was building, not chronometers, but clocks which reflected all the cosmic movements and were at the same time synchronously functioning astrological diagrams."[94]

Baldewin's, Brahe's, and Bürgi's time-keeping talents modeled the heavens for astronomical and astrological purposes. The astrological functions of these clocks were probably useful in obtaining patronage that not only furthered astrological studies, but more importantly for the history of science and technology, supported astronomy and horology.[95]

There is also something else going on that has been less appreciated in the history of timekeeping—the use of multiple clocks as checks on each other. Clocks made before the invention of the pendulum have a reputation for being highly inaccurate, but this contrasts with the accuracy and precision of Brahe's observations even as he complained of the poor performance of clocks.[96] This seeming contradiction is resolved by considering that patrons of science encouraged the accumulation of multiple clocks. This was not just a matter of ostentatious display in Kunstkammers but created reasonably accurate timekeeping. Having multiple clocks allows for comparison. Such comparison allows the elimination of poorly performing clocks, and also the calculation of time. As pointed out earlier, Charles V obsessed over synchronizing his clocks. His son, Philip II, had the same desire, which intimidated his valet, Jehan l'Hermite. Jehan l'Hermite was responsible for maintaining the clocks and scientific instruments in his patron's collection, and about this, he complained, "Je me trouvay encore embrouillé avec grande quantité d'horloges et plusieurs instruments mathematiques" (I found myself confused with the great number of clocks and many mathematical instruments).[97] Such practices are the foundation of the current reliance on suites of atomic clocks for the determination of UTC (see chapters 7 and 8), as well as the use of multiple time sources in the computer-time-synchronization protocol that is known as Network Time Protocol (NTP). By using multiple clocks, one can hedge against the malfunctioning of one clock.

Brahe's discussion of clocks in his discussion of the wall quadrant in the 1602 edition of his *Astronomiae Instuaratae Mechanica* demonstrates how Brahe (and likely Rudolf) obtained accuracy in time measurement. In a passage added to the new edition of this book, Brahe wrote "duo ad minimum esse opertet horologia eiuscemodi, quo unum alterum emender" (at minimum choose two of this kind of clock, that one may correct the other).[98] Indeed, on the page before the discussion of the wall quadrant, in both editions of the book I have discussed, the drawing of the wall quadrant includes a person checking two side-by-side clocks.[99]

Patronage, Collecting, and the Emergence of an Interest in Accuracy

Just as with current practice, which recognizes that no single atomic clock is entirely accurate and that, therefore, time should be calculated from a number of clocks with the performance of each individual clock noted, Brahe used patronage to pioneer the use of a suite of clocks to determine the correct time, which fit with the attention he gave to analyzing errors.[100] By the same token, Rudolf II likely did not accumulate celestial globe clocks just to demonstrate that he was a major collector, but to use them together to know the time and to check their performance. This was not a matter of moving toward a single timescale, however, but instead of reflecting the ability to compare clocks and convert between different hour systems.

There is evidence that Rudolf II could ascertain poor performance in clocks as a result. In 1584, Georg Roll and Johannes Reinhold sold two celestial globe clocks: one to Rudolf II for 1200 thalers, and the other to Rudolf's brother Ernst for 1500 thalers. Unfortunately for the makers of the globe clocks, the one acquired by Rudolf had a defect. Hans Popp, Rudolf's valet, wrote that Ernst had a better globe clock than the emperor, and likened the emperor's clock to a rotten fish.[101] Rudolf was outraged. He quickly learned of the price difference between his clock and that which Ernst had bought, and he believed that he had been sold a globe clock of lesser quality than the one sold to his brother. Roll asserted that the two clocks were identical. This seemed to be a mistake on Roll's part, for when he was asked if the emperor had the better of the two clocks, Roll truthfully responded "nein," because, after all, the two clocks were the same. Rudolf wrote to Augsburg to have Georg Roll put in prison as punishment for this. In July 1585, the leaders of Augsburg informed Rudolf that they had arrested Roll. Soon after, Roll wrote directly to Rudolf stating that the two globes were equal in value, and that therefore Rudolf had gotten a much better deal than his brother. Later that month, Rudolf replied, confirming his receipt of the letters from Augsburg and Roll, and replied that he had doubts about Roll's account and hoped that Roll would be wiser now. After Ernst died, Rudolf acquired what he thought was the "better" of the two clocks.[102]

While it is not known exactly what the defect in Rudolf's clock was, it is likely that it was that this particular clock stood out as a poor performer when compared to Rudolf II's other clocks. Part of Rudolf II's reputation was not only due to his accumulation of clocks, but his ability to judge them because of his vast collection. This had an additional effect of raising the stakes for anyone who gave Rudolf II a clock—it had to perform well or else it would embarrass the giver.

Not only did individuals serve as patrons of gifts, but cities did as well.[103] Gifts were meant to define relationships, but in so doing, they did much more. In the case of clocks, gift clocks became a nexus of patronage of craft skill, presentations of the patron's identity, and anticipated desires on the part of the recipient. When accumulated in numbers, gift clocks enhanced their recipient's ability to achieve reliable timekeeping. In this context, however many clocks one had, an additional well-made clock would be welcomed. All of these ended up shaping clocks and the development of horology—particularly the interface between the movement of celestial objects, the representation of time, the promotion of multiple hour systems, and the ability to convert between them.

A gift could have unforeseen consequences. The Duke of Brunswick gave Emperor Rudolf II a clock by Jost Bürgi that included a representation of the movement of the planets.[104] By graphically representing the movement of the planets, it visually confirmed the inaccuracy of Copernicus's theory of circular (as opposed to elliptical) planetary motion. Combined with Tycho Brahe's observations, Bürgi's mathematical innovations in decimal fractions and logarithms,[105] and Bürgi's own collaborations with Johannes Kepler, this clock spurred Kepler's efforts to develop new laws of planetary motion.

It is an old adage in philosophy that the question one asks determines the answer one is given. A corollary would be that the question one asks shapes the means one uses to determine the answer. What one wants to know about time shapes how a clock is designed. The Habsburgs were obsessed with astrology and cosmopolitan displays of multiple hour systems. Their timekeeping sought to know the astrological significance of a particular moment. Clocks played a critical role in knowing the precise timing of transits.[106] The Habsburgs also governed a polity in which there were numerous independent cities that asserted their own identities and their own timescales, and yet they were tied together by the empire. Complementing astrology, Habsburg clocks allowed the conversion between the popular timescales found within their realms. When a Habsburg gave a clock, the clock included the two functions of astrological reckoning and timescale conversion. The questions about time the Habsburgs pushed the recipients of their clocks to ask reflected Habsburg interests.

Cosmopolitanism in timekeeping fit a particular governing philosophy that allowed for decentralized administration, but Habsburg colonization waned in favor of British and French colonial efforts, and these efforts were tied to a different time-keeping philosophy that closely linked time and space. The seeds of this philosophy emerged within the Habsburg context and can be most clearly seen in how Regiomontanus imagined the relationship of the globe and timekeeping, but as the next chapter documents, Regiomontanus seemed comfortable with Habsburg cosmopolitanism.

Notes

1. Bede, *Reckoning of Time*, 13.
2. See Schukowski, *Wunderuhren*.
3. Koenigsberger, *Habsburgs and Europe*, 1.
4. Scheicher, *Kunst- und Wunderkammern*, 187; see also Garcia-Diego, *Juanelo Turriano*, chapter 5.
5. McGuigan, *Habsburgs*, 66.
6. Mignet, *Charles-Quint*, 248–49.
7. Vehse, *Memoirs of the Court*, 169.
8. van Cleempoel, "Philip II's Escorial," 109.
9. Goodman, *Power and Penury*; Rodríguez-Salgado, "Court of Philip II," 211.
10. Goodman, *Power and Penury*, 20.
11. Thomas, *Decline of Magic*, 285.
12. Thorndike, "True Place of Astrology"; White, *Medieval Religion and Technology*, 297–315.
13. Gustafsson, "Conglomerate State."
14. Elliott, "Composite Monarchies."
15. Press, "Imperial Court of the Habsburgs," 294.
16. Brandi, *Emperor Charles V*, 99.
17. Moraw, "Court of the German King," 112.
18. Hamburger, "Introduction," 5.
19. Evans, *Making of the Habsburg Monarchy*, 151.
20. Koenigsberger, *Early Modern Europe*, 42.
21. Koenigsberger, *Early Modern Europe*, 42.
22. Evans, "Austrian Habsburgs," 121.
23. Evans, *Making of the Habsburg Monarchy*, 22.
24. McGuigan, *Habsburgs*, 98.
25. Elliott, "Composite Monarchies," 64.
26. Findlen, *Possessing Nature*, 3.
27. Findlen, *Possessing Nature*, chapter 5.
28. Rose, *Italian Renaissance*, 2 and 26.
29. Turner, "Cabinet of Experimental Philosophy," 296–301; Goodman, *Power and Penury*, 54–55.
30. North, *The Ambassadors' Secret*, 73.
31. Quiccheberg, *Inscriptiones*, 5.
32. Schlosser, *Die Kunst- und Wunderkammern*, 30.
33. Kaufmann, *Variations*, 108–9; Kaufmann,"Remarks," 22; Fučíková, "Belvedere in Prague," 276.
34. Marshall, *Mercurial Emperor*, 85.
35. Findlen, *Possessing Nature*, 129.
36. Findlen, *Possessing Nature*, 131.
37. Findlen, *Possessing Nature*, 132.
38. Apian, *Astronomicum Caesareum*.
39. Machiavelli, *The Prince*, 5.
40. Heal, *Power of Gifts*, 43–46.
41. Menzhausen, "Elector Augustus' *Kunstkammer*," 96–97.
42. See chapter 7; also Gaulke, "The First European Observatory," 95.
43. Moran, "Wilhelm IV," 79.

44. Regiomontanus and Schöner, *Scripta Clarissimi Mathmatici.*
45. Apian, *Astronomicum Caesareum.*
46. Schöner and Wilhelm IV, "Beobachtungen und Aufzeichnungen."
47. See Biagioli, "Galileo's System of Patronage," 17.
48. Moran, "German Prince-Practitioners," 269–70.
49. e.g., see Ramelli, *Le diverse et artificiose machine*; Veranzio, *Machinae Novae.*
50. Moran, "Machines," 22.
51. See H. Lloyd, *Some Outstanding Clocks*, 46–57.
52. Gaulke, "The First European Observatory," 88.
53. Gaulke, "The First European Observatory," 88.
54. Moran, "Wilhelm IV," 76; Mathematisch-Physikalischer Salon, *Uhren*, 106–9.
55. Bedini, "Evolution of Science Museums," 11.
56. Schillinger, "Saxon Scientific Instrument-Making Skills."
57. Schillinger, "Saxon Scientific Instrument-Making Skills," 279.
58. Schöner, *Gnomonice*, dedication.
59. Moran, "German Prince-Practitioners," 255.
60. Bedini, *Patrons, Artisans and Instruments*, xiii.
61. Ashworth, "The Habsburg Circle," 137.
62. White, *Medieval Religion and Technology,* 86 and 124.
63. White, *Medieval Religion and Technology*, 198.
64. Biagioli, "Galileo's System of Patronage," 3.
65. Moran, "Wilhelm IV of Hesse-Kassel," 67.
66. Moran, "Wilhelm IV of Hesse-Kassel," 74.
67. Seed, *Ceremonies of Possession*, 121–31.
68. Harley, "Silences and Secrecy," 59.
69. Black, *Maps and Politics*, 122–25.
70. Wallis and Robinson, *Cartographical Innovations*, 93.
71. Black, *Maps and Politics*, 122.
72. Mukerji, "Visual Language in Science," 31.
73. Huebener et al., "Time and Globalization," 245; Wallis and Robinson, *Cartographical Innovations.*
74. Harley, "Rereading the Maps," 523–34.
75. Akerman, "Introduction," 1–2.
76. Goodman, *Power and Penury.* 54–72.
77. Wallis and Robinson, *Cartographical Innovations*, 178.
78. Turnbull, "Cartography and Science."
79. Mukerji, "Visual Language in Science."
80. Ziegler, *Regiomontanus*, 13.
81. Heilbron, *Sun in the Church*, 7.
82. Berggren and Jones, "Introduction," 3.
83. Ptolemy, *Almagest*, II.6.
84. Ptolemy, *Geography*, book I.
85. Wallis and Robinson, *Cartographical Innovations*, 181.
86. Berggren and Jones, "Introduction," 24–25; Ptolemy, *Geography*, I.14.
87. Doppelmayr, *Historiche Nachricht*, 9 and note d.
88. Brahe, *Tychonis Brahe* (1598 edition).
89. Brahe, *Tychonis Brahe* (1598 edition), 2.
90. Brahe, *Tychonis Brahe* (1598 edition), 3.
91. Brahe, *Tychonis Brahe* (1598 edition), 4.
92. Gaulke, "The First European Observatory," 92.

93. Staudacher, *Jost Bürgi*, 88–91.
94. Maurice, "On Innovation," 101.
95. Rose, *Italian Renaissance of Mathematics*, 15.
96. Brahe, *Tychonis Brahe* (1602 edition), 148–49.
97. l'Hermite, *Le Passetemps*, 173.
98. Brahe, *Tychonis Brahe* (1602 edition), 26.
99. Brahe, *Tychonis Brahe* (1598 edition), 21; *Tychonis Brahe* (1602 edition), 25.
100. Chapman, "Tycho Brahe," 74; Wolfschmidt, "Observatories and Instruments," 210.
101. Bobinger, *Kunstuhrmacher in Alt-Augsburg*, 34.
102. Bobinger, *Kunstuhrmacher in Alt-Augsburg*, 34–35.
103. See Groebner, *Liquid Assets*.
104. Bukovinská, "*Kunstkammer* of Rudolf II," 203.
105. Staudacher, *Jost Bürgi*, 178–79, 215.
106. White, *Medieval Religion and Technology*, 301.

Chapter 4

Time Pluralism and Scientists' Choices

> Tycho-Brahé avoit essayé tous les moyens qu'il s'estoit pû imaginer, pour mesurer éxactement le temps, soit par les clepsydres d'eau, de mercure, & de diverses autres liqueurs ; soit par d'autres manieres d'horloges qu'il avoit fait faire sur différens principes. Mais après estre épuisé sur ce sujet, il fut oblige d'en revenir aux horloges ordinaires ...(In order to measure time exactly, Tycho Brahe tried every means he could imagine—either by clepsydrae of water, mercury, and various other liquids, or by means of clocks that he had made based on different principles. But after exhausting the subject, he was forced to revert to ordinary clocks ...)[1]
>
> —Giovanni Dominico Cassini, *Recueil d'observations*

Cosmopolitan time pluralism in timekeeping poses two challenges: choosing an hour system to use at any given moment and converting between hour systems. In Habsburg domains approaches to timekeeping did not emphasize uniformity. There was a political preference for maintaining and displaying multiple hour systems. In the same period, there was a burgeoning interest in astronomy and cartography in which a patchwork of local timescales posed a challenge in the communication of observational data from one locale to another. Moreover, the invention of the printing press allowed for the duplication and distribution of observational knowledge at an unprecedented scale. How did scholars and handle this? It seems they made choices and learned to convert between timescales.

Both UTC and Habsburg timekeeping share an interest in widely distributing time-keeping knowledge. Today, it is through satellite signals and computer networks, but in the Renaissance, the technological innovation for distributing ideas about hour systems and converting between them was tied to the use of the printing press.

Sometime after the Renaissance astronomer, Regiomontanus, moved to Nuremberg in 1471, the city leaders asked him to use his skill to give the city's system of counting hours a stronger foundation in celestial cycles than it previously had. Being from Königsberg and having spent a lot of time in cosmopolitan timekeeping circles, Regiomonatanus was probably familiar with Nuremberg's system of different counts of hour for the daylight and nighttime hours. His complicity with the request suggests that he did not mind that the public timescale was different from the time system he used in his research. What makes Regiomontanus and his protégé, Bernard Walther, of interest for understanding time pluralism is that they were scientists working in a context of time pluralism and even enabling it. The textual record Regiomontanus left behind provides insights into his choices.

This textual record does more than document Regiomontanus's choices, however. His use of the printing press to distribute his ideas shifted the access to scientific knowledge away from a small group of astronomers to a much wider audience. This fed a process by which timekeeping and astronomy became linked with cartography and eventually with the European imagination of how borders and colonies were geographically defined.

Regiomontanus's Cosmopolitan Background

Regiomontanus was born in 1436. He was a mathematical prodigy who attracted the attention of Georg von Peuerbach. Peuerbach was the leading European astronomer of his time and lectured on astronomy at the University of Vienna. There, he served as astrologer to King Ladislas V of Bohemia and Hungary, a 14-year-old monarch who spent most of his time in Prague and Vienna. Ladislas's residence in Vienna not only allowed Peuerbach to teach at the university, but it also put him into contact with Ladislas's guardian and the regent to the Hungarian and Bohemian throne—the Holy Roman Emperor Frederick III.

Regiomontanus studied at the University of Vienna, where Peuerbach became his teacher and mentor. He soon started to assist Peuerbach in working on a commentary on the *Almagest* of Ptolemy, and it was Regiomontanus who completed this work after Peuerbach's death. Ptolemy's work had become known to Western European astronomers shortly before. Using Ptolemy as well as Sacrobosco's work on spherical astronomy,[2] Regiomontanus made scientific instruments for Peuerbach.

In 1460, Regiomontanus met Cardinal Basilios Bessarion, who at the time was the papal legate to Vienna. Bessarion made significant contributions to the advancement of mathematics and science because of his knowledge of Greek. This gave him access to texts that had not previously been translated into Latin, such as works by Archimedes, Aristotle, and Ptolemy. In Bologna, he supervised the construction of a public astronomical clock.[3] Bessarion was so impressed with Regiomontanus that he made him his personal secretary. Together they would write a work on the error in the Julian calendar, *De Errore Paschalis*,[4] which contributed to the laborious process of calendar reform that produced the Gregorian calendar.[5]

Regiomontanus left Bessarion to work on the establishment of a university at Pressburg to rival that of Bologna. This was the idea of King Matthias Corvinus of Hungary and Johann Vitéz, Archbishop of Gran. Corvinus was an enthusiastic supporter of books, the arts, and science, as well as a rival to the Habsburgs. Vitéz, like Bessarion, was a church official with an interest in mathematics and astronomy. Like Regiomontanus, he was also a protégé of Peuerbach. Regiomontanus's duties were to teach and to produce scientific instruments and tables for the archbishop. Corvinus and Vitéz then asked Regiomontanus to move to Buda to supervise Corvinus's library and to make scientific instruments.[6] Vitéz was against Corvinus's war against Bohemia and was at the center of a conspiracy to undermine the war and engineer Casimir to become king of Bohemia and even Hungary.[7] This led to Vitéz's imprisonment by Corvinus in 1471, and the closing of the university at Pressburg. In these circumstances, Regiomontanus decided to relocate.

His chosen destination was Nuremberg. He shared his reasons for this choice in a letter to the mathematician Christian Roder of Erfurt: Nuremberg produced high-quality astronomical instruments designed by astronomers, it had a large number of "studious men," and it played a central role in trade in Europe.[8] The technology of Gutenberg's printing press had arrived there by 1470,[9] and the city became a major publishing center.[10] In sum, it had a combination of technical and scientific expertise unrivaled in Europe at the time.[11] Nuremberg's support for astronomical observation and the dissemination of results was unequaled, but its cosmopolitanism and pride in its own hour system worked against the standardization that current observational science demands.

Patronage and Bernhard Walther

As discussed in the last chapter, it was a common practice for scientists to seek patronage by giving gifts.[12] As an instrument maker, in 1462 Regiomontanus made and gave an astrolabe to his patron, Cardinal

Bessarion.[13] Regiomontanus also made a folding sundial with the image of Pope Paul II, which is now in the German National Museum in Nuremberg. Regiomontanus made automata, with his skill likely augmented by legend. For instance, he supposedly made a mechanical fly that he gave to the Emperor Frederick III, as well as a mechanical eagle that escorted the emperor on his entry into Nuremberg.[14]

Regiomontanus's patrons shaped his science and the correspondence he carried out with other astronomers, but once he moved to Nuremberg, his patron there, Bernhard Walther, was also his student and interlocutor. This was a special relationship because of its duration and the extent to which Walther was both a reliable patron and a disciple. Walther was a merchant who used his wealth to become one of the most important astronomical observers of his time. He had arrived in Nuremberg four years before Regiomontanus, and upon the latter's arrival, Walther provided support for a printing press and astronomical instruments. In this respect, Regiomontanus and Walther were pioneers in melding material technology, literary technology, and social technology—a combination that would prove significant in cultural construction of facts and their distribution in Enlightenment science.[15] In return, Regiomontanus mentored and trained Walther so that the two eventually became equal partners in their astronomical work—so much so that, after Regiomontanus's death in 1476, Walther continued Regiomontanus's program and built upon it with his own innovations,[16] compiling a thirty-year record of observations that would be the most accurate until those of Tycho Brahe.

Public Service

Patronage shaped the timescales clocks displayed, but the use of clocks was not limited to the concerns of ruling families. As discussed in chapter 2, guilds referred to clocks to regulate work hours; city governments regulated the opening and closing of markets by means of the clock; and religious institutions continued to use temporal/canonical hours to structure the liturgical cycles of each day. In astronomy, the measurement of time was crucial, but modern astronomy emerged in the context of time pluralism. Not only did observers have to manage their local time systems, but in their correspondence, they often had to convert between the different hour systems used by observers in different locales.

These different ways of counting hours were, in part, based on different ways of determining when the day began. The concept of the "day" seems like a natural one, although that is misleading. The definition of the day and determination of when a day begins is one of the most culturally variable and long-contested dimensions of timekeeping.[17] A feature of the cosmopolitan timekeeping found in many parts of Renaissance and early modern Europe was different starting points for the hours of the day, such as sundown in Italian hours, or midnight in other systems. Choosing a timescale sometimes meant choosing a definition of the day, although not necessarily so. Such was the case with Ptolemy in his *Almagest*. For him the day started when the Sun crossed the meridian,[18] but his count of hours began at sunrise.[19]

Knowledge of different systems of day reckoning was part of being educated about time and found within medieval pedagogies about timekeeping. In the seventh century, Isidore of Seville discussed this.[20] Similar passages can be found in Bede's eighth-century work on time,[21] and the *Disputatio Puerorum* attributed to Alcuin.[22] Expanded versions of this classification of day systems persisted, such as in a poem reproduced in Alsted's seventeenth-century *Scientiarum Omnium Encyclopaediae.*

> Atticus occasum spectat, Babylonius ortum:
> Nox media Ausoniis, media at lux perplacet Umbris:
> Graeci manè diem incipiebant solis ab ortu:
> Vespere Iudai: Scrutantes sidera coeli,
> Dum sol in medio fulget nitidissimus orbe :
> Christicolae incipient medio sub tempore noctis.
> (Athenians considered the setting, Babylonians, the rising:
> Midnight for the Italians, but on the contrary, the Umbrians reckon it with the middle of the day:
> For the Greeks, they begin early in the morning from sunrise:
> Sundown for the Jews: scrutinizing the stars of the heavens,
> While the sun at midday shines brightly on the earth:
> The Christians begin in the middle of the night.)[23]

Alsted's reference to the verse does not claim it as his own, but represents it, like later authors, as a verse of unknown authorship. According to Paolo Rossi, Alsted's encyclopedia was "one of the most widely circulated works of seventeenth century European culture,"[24] and this rhyme was likely widely shared pedagogical doggerel to help students to remember differences in the start of the day. Moreover, the verse continues to be cited through the nineteenth century.[25] It was only with the standardization of global timescales, in the late nineteenth century, that the verse fell into disuse.

Nuremberg

Nuremberg was a center for the manufacture of scientific instruments. As a result of the circulation of merchants peddling its goods, it became a center for communication, and a meeting place of different hour systems and definitions of the day as indicated by broadsides that juxtaposed these systems.[26] As a node in networks of information, it led the development of geography and cartography even though it was not directly involved in exploration and colonial expansion.[27] In Nuremberg, scientists linked astronomy, astrology, geography, and cartography. For instance, in his preface to Johannes Schöner's book on astrology and nativities, the Lutheran leader Philipp Melancthon describes astrology as the "fons est Geographiae" (It is the source/spring of geography).[28]

Nuremberg was famous for its distinctive timescale called Nuremberg hours or great hours. As described in chapter 1, in this system, hours were of equal length, but a different count of hours was kept for daytime and nighttime hours. As a result, the number of daytime and nighttime hours varied with the season and were only equal in number at the equinoxes. For instance, at the winter solstice, eight o'clock in the day would be followed by one o'clock at night, and sixteen o'clock at night would be followed by one o'clock in the day.[29] Each count also shifted in relationship to solar noon as the timing of sunrise and sunset changed. As Johann Georg Keyssler, a traveler to Nuremberg in 1756–57, described it: "The principal clocks in this city strike the hours from one to 12 progressively from the rising of the sun, and begin again after sunset, which at first puzzles strangers; but by the table of the aequation of time, published here [in Nuremberg], this may be reduced to the common method of calculating time."[30]

Regensburg, Schwabach, Windsheim, and Rothenburg also used Nuremberg's hour system.[31] Even though this timescale would come to be displayed on table clocks during the sixteenth and seventeenth centuries, in Nuremberg human bell ringers rather than clocks signaled these hours.[32] William Smith, a late sixteenth-century traveler to the city, documented this: the bell at St. Sebald's struck the hours, then upon hearing this, the hours were struck at the Innerlauffer Tower in the east part of the city, and then St. Lawrence's church in the south of the city would strike the hours by hand, and finally, the White Tower in the west struck the hours.[33]

The earliest evidence of these great hours as they were called, is from 1374 in a birth record.[34] By 1452, they were used to represent the times of the workday. Surviving from that year is a builder's book that lists the hours of work based on the amount of daylight, and reckoned in these great hours; for instance, "Item: wenn es den tag xiiij slecht, so sol man zu morgens an die arbeit gen zu einer or auf den tag, und zu der suppen zu iij oren, und zu siben oren auch wider davon, unde die abent zeit davon zu zehen oren, und zu nacht ein or gen nacht Darvon" (Item: when the day is fourteen hours long, so should one go to work at one of the day, and to a meal at the third hour, and again at the seventh hour, and [work] until the tenth hour in the evening, and from the hour of one in the night go from work).[35] In contrast, when the daylight was ten hours in length, the first meal was at the second hour, and the second meal was from midday until the fifth hour, after which work continued until nightfall.[36]

As Dohrn-van Rossum points out, such an hour system needs astronomical expertise to get the counts of hours to match the duration of day and night.[37] While Regiomontanus did not create great hours for the city as some folklore suggests, his work contributed to their definition.[38] In his *Kalendarium* of 1474, Regiomontanus provided a table on the length of the day as represented in equinoctial hours.[39] Regiomontanus's protégé Bernhard Walther used his observations on the day length for Nuremberg to improve the reckoning of this timescale.[40] Such improvements served as the basis for clockmakers constructing dials that could represent Nuremberg hours in relationship to Italian and small clock hours, such as is found on many sixteenth- and seventeenth-century clocks.

Regiomontanus

Regiomontanus and his patron and protégé Walther used scientific instruments for observation to shift European astronomy away from relying solely on tables toward empirical research. Building on Ptolemy's *Geography*, their work also encouraged a consciousness of the globe that not only envisioned the heavens as a celestial sphere, but the Earth as a globe on which different positions led to different perceptions of the heavens. In addition to his scientific research, Regiomontanus became a publisher. This was an innovation in the mass production and distribution of scientific knowledge. The contributions of Regiomontanus and Walther are less in terms of innovations in astronomical theory, and more in terms of laying the foundation for the distribution of scientific techniques and methods through the use of the printing press—what has been described by Jan Golinski as the "laborious processes of 'standardization.'"[41] As such, they form part of the transition from the ethos of cosmopolitanism found in locations of Habsburg Europe to the ethos of global standardization in current time metrology.

In contrast to twenty-first-century time metrology, which seeks a single publicly disseminated timescale, Regiomontanus and Walther cultivated multiple timescales. They supported the city's having different counts of hours for daytime and nighttime, but in the records of their observations, Regiomontanus and Walther adopted a timescale different from that of the city. Moreover, in Regiomontanus's publications, he discussed how to convert between different systems of counting hours.

The Ptolemaic Challenge

In the fifteenth century, the newly available translations and discussions of Ptolemy's *Almagest* in Latin influenced Renaissance astronomers. Ptolemy is known for his theory of a geocentric universe, but his work also contains a great deal of discussion about calculating the location of positions on the Earth and its size. In their history of cartography, Helen Wallis and Arthur Robinson argue that such information improved navigational charts.[42] Regiomontanus's relationship to this work was particularly intense since he had worked with Georg von Peuerbach on a commentary on Ptolemy's *Almagest*.[43]

Just as influential as Ptolemy's works were the Alfonsine tables.[44] These tables shaped late medieval and Renaissance astronomy.[45] They were commissioned by King Alfonso X of Castile in the mid-thirteenth century, and were widely circulated in manuscript form, then printed in the fifteenth century. The tables charted the movement of celestial objects in the skies of Toledo, and by means of simple conversions based on latitude and longitude, astronomers and astrologers adapted the tables to other locations. Their popularity among astrologers and astronomers became the basis for many other tables. The Alfonsine tables had many discrepancies from observed phenomena,[46] but since in medieval astrology there was a greater emphasis on calculations based on tables rather than on direct observation, such differences were not noted until the fifteenth century when many new observational tools and techniques came to be applied. Again, Regiomontanus had a close connection to these efforts, and he saw one of his goals as correcting the inaccuracies of such existing astronomical tables.[47]

The importance of Ptolemy and the Alfonsine tables posed a cognitive challenge for astronomers, however. Ptolemy's observations and the Alfonsine tables both define solar days and equinoctial hours from when the Sun crosses the meridian—what is colloquially called *solar noon*. This practice persisted into the twentieth century among astronomers. In contrast, Ptolemy's civil calendar reckoning defines the day as beginning with sunrise. Consequently, the calendar dates and astronomical days in Ptolemy do not match.[48] Using Ptolemy and the Alfonsine tables, Regiomontanus and Walther had to reconcile calendar dates with astronomical timekeeping. They also had to choose how to count hours and when the day began. In defining the day, they actually differed: Regiomontanus's observational notes use equinoctial hours beginning after the Sun passed the meridian,[49] and Walther used these hours starting at midnight. In effect, their observations occurred in a context of time pluralism, thereby forcing them to make choices.

In addition, Walther provides the earliest explicit application of clockwork in astronomical observation, but only to measure duration, not to determine the time of day. Tycho Brahe described Walther's method for measuring the relationship of Mercury's rising to the Sun. In Brahe's understanding, the moment Mercury rose, Walther attached a weight to a clock on which the hour gear had fifty-six teeth. He then counted the number of teeth that "*praeterlissent*" (passed before) until the Sun rose. In the case of this particular observation of Mercury, Walther counted one full revolution of the dial plus thirty-five teeth, which he calculated as an interval of 1 hour and 37 minutes, yet in other observations, Walther used the time of the transit of the meridian.[50] His approach was not so much to determine the time of day of a celestial event as to measure the duration between two events.[51]

Despite their involvement in Nuremberg's distinctive timescale, and despite the dominance of canonical hours for much daily timekeeping throughout Europe, for their observations, Regiomontanus and Walther preferred the use of hours of equal length. Building on Regiomontanus's instructions for determining equinoctial hours,[52] one of his admirers, Johannes Schöner, by 1539, wrote about how to determine astronomical hours—these are equinoctial hours counted from when the Sun crosses the local meridian,[53] and in his work on spherical instruments, Schöner describes the use of astronomical hours to determine the time of the rising of fixed stars.[54]

Regiomontanus's and Walther's use of equinoctial hours to determine the position of the planets, Sun, and Moon, and to represent the length of day suggests the shift away from privileging the temporal hours and Nuremberg hours. When taken as a whole, much of the information Regiomontanus published served one of two common purposes: knowing the moveable feasts in the Church calendar or knowing celestial cycles for purposes of astrological calculations. There was also a less obvious, geodesic purpose. Early in his *Ephemerides*, he provides instructions for how to convert the information in the tables to one's specific meridian.[55] With the *Ephemerides* and a knowledge of longitude and latitude, one could predict what would be visible in the sky.

Scientific Imagination

Clocks and astronomical tables are cognitive tools that mediate between the expert knowledge of their designers and their users, thereby allowing users to access the knowledge of experts. Because of his expert knowledge and his use of printing to distribute his knowledge, Regiomontanus played a key role in the imagination of timekeeping. One documented consequence of the impact of his work through publishing is the influence his *Epitome of Ptolemy's Almagest* had on Copernicus.[56] This shift from a science built on personal correspondence to a science built on the mass distribution of printed matter is one of Regiomontanus's most important innovations. He pioneered the mass-distributed cognitive tool—an easily reproduced and distributed package of information. By providing techniques to convert his tables to any location on the globe, he combined the mass distribution of his tools with the global usefulness of these same tools. In this way, Regiomontanus made a significant contribution to the global imagination critical in the knowledge used in European imperialism even though he died seventeen years before Columbus's first voyage. Since Regiomontanus used Ptolemy's methods of counting time, Regiomontanus's count of equinoctial hours began at noon. The mass distribution of his works influenced others to adopt his system. This was a slow process, but it eventually overcame the challenge of converting between different locally based timescales by adopting a single, uniform timescale that only varied from location to location due to longitude.

In addition to the distribution of his works subtly encouraging observers to adopt equinoctial hours starting at noon, Regiomontanus provided a model for the conversion of data observed with different timescales. In a letter to Giovanni Bianchini, the astronomer of the Duke of Ferrara, Regiomontanus described a question put to him by Master Christoforo, who had observed a star he did not know on 5 April 1463. Christoforo described the star as crossing the meridian of Ferrara at the second hour and twenty-fifth minute of the clock, and specified the location as latitude 44 degrees, 45 minutes, and 4 seconds, and the longitude as 32 degrees. The altitude of the star was 39 degrees 16 minutes.[57] In this record, the second hour is not 2:00 AM, but the second hour in Italian hours, which were counted starting about a half hour after sunset. To solve the problem, Regiomontanus had to determine when sundown was in Christoforo's location according to equinoctial as opposed to Italian hours. He determined the declination of the setting sun for that date and place and then converted the degrees of declination into a time difference from the equinox. Using this, he ascertained that sunset was at 6:38 PM, and then determining when the clock would have started for Italian hours, the count of which started after sunset—he calculated that the star in question had crossed the meridian at 9:03.[58] In the notes on his calculations that he included with the letter, it becomes clear that his calculations also involved converting astronomical information for Venice into comparable information for Ferrara.[59]

This suggests that, in creating problems for reconciling observations from different locations, time pluralism created problems that demanded solutions. These solutions involved developing a vision of the Earth that could determine what was visible in the sky at any latitude or longitude based on calculations and converting between different timescales used for observations. The goal was to move science away from its local limits to create a global vision so that observational science could be carried out anywhere yet be meaningful everywhere—a crucial step toward global standards.

Geodesy

This reveals another dimension of Regiomontanus's work and vision—the study of geodesy. Geodesy is the study of the size, shape, motion, and orientation of the Earth. For Regiomontanus, this involved linking ideas of time to cartography. Regiomontanus was interested in the work of Paolo dal Pozzo Toscanelli. Toscanelli was a Renaissance mathematician and humanist who won the respect of his contemporaries. He met Regiomontanus when Regiomontanus accompanied his patron, Cardinal Bessarion, to Rome in 1461.[60] Toscanelli's astronomical observations had been conducted with the famed architect Leon Battista Alberti. Both combined astronomy and mathematics to map space. It was from this sort of approach that Toscanelli estimated the circumference of the Earth at the equator and created his map of the world—a copy of which Columbus acquired. In writing about technology in Philip II's rule of Spain, David Goodman argues that this use of astronomy to map the Earth was increasingly important in an age of colonial expansion and global imperial claims.[61]

In Regiomontanus's introduction to the 1474 edition of his *Ephemerides*, he outlines its uses, some of which are related to geodesy. These included the following features. He said it would aid in knowing the golden number of the year as well as the dominical letter of each day of the year—information used to calculate when Easter and its associated moveable feast days would occur. It would allow the prediction of eclipses and chart the retrograde motion of the planets. It would indicate the relative positions of the Sun, Moon, and planets to the meridian in terms of degrees and allowed the user to determine the motion of any planet at any time of day.[62] In fact, the *Ephemerides* included information for every day between 1475 and 1506 for the position of the Sun, Moon, and planets, as well as moon phases, planetary conjunctions, and their effects on weather. This was also information that could be used as a navigational aid for determining one's position on the open sea.[63]

The calendar he published the same year provided very similar information, but with more discussion of how to determine the timing of Easter and how to use the table to determine the time of day a conjunction occurs. He also offered the length of daylight during the year, although he charted this according to zodiac signs rather than months. To aid users in determining the time, he described how to make a horizontal sundial and how to determine equinoctial hours and the temporal hours.[64] In fact, his *Ephemerides* became extremely popular and were reproduced after his death. These works distributed a means for imagining Earth in terms of coordinates in relationship to time. They also encouraged the adoption of a single time system to think about the relationship of different locations on the globe. In effect, they were a step towards the development of the coordinate timescale we use today.

Several famous explorers used Regiomontanus's *Ephemerides*, including Columbus, who took with him a copy of Regiomontanus's *Ephemerides* on voyages.[65] Regiomontanus's protégé Martin Behaim[66] became an important figure in the court of King João II of Portugal and played a role in shaping Portuguese exploration and navigation.[67] Indeed, Behaim's globe, made in 1492, demonstrated the ease with which one could sail westward to reach the East Indies. Because Behaim underestimated the size of the Earth, he thought that if one sailed west from Europe, the East Indies were closer than they really are. His globe did not represent North or South America and showed nothing but ocean between Europe and Japan.[68] He apparently constructed the globe without knowing that Columbus had secured support in Spain to sail west.[69]

Columbus, like Behaim, underestimated the Earth's circumference at the equator. It was on the basis of Toscanelli's map that Columbus thought he had reached the Indies by sailing westward from Europe. It was on the basis of Regiomontanus's *Ephemerides* that Columbus predicted the lunar eclipse of 1504 when he was in Jamaica. But Columbus also got his prediction wrong because he did not take into account that Regiomontanus's tables were set to Nuremberg, not to Columbus's port of departure, Cadiz.[70]

Regiomontanus's and Walther's use of instruments contributed to the development of techniques of surveying in order to map territories and make claims. Early maps of the Americas bear evidence of the use of cartography in European claims to territories.[71] Regiomontanus work laid the foundation for later European colonialism and its distinctive feature of mapping territories to divide them. He cannot be blamed for the consequences of European expansion, but the form that expansion took in terms of navigation, surveying, and documentation of borders would not have been possible without Regiomontanus's and Walther's work.

Regiomontanus's global vision depended on uniform, equinoctial hours even as he used his knowledge to support timescale cosmopolitanism. The use of homogenous, empty time attributed to modernity and viewed as necessary for the emergence of nationalism by Benedict Anderson was built on Regiomontanus's global scientific vision and premised on his divergence from the timekeeping associated with the Church and with imperial rule during his lifetime.[72]

A Paradox

One might think that clocks with multiple hour systems would be a hindrance to scientific advances. Yet, when it comes to astronomy, one of the sciences most closely related to horology, the great advances in this period are associated with cities in the Holy Roman Empire, and hence cities that embraced temporal pluralism. Regiomontanus and Walther chose Nuremberg to be their base of operations. Copernicus studied in Krakow and Bologna. If surviving masterpiece clocks are an indication of requirements for clockmaking in Krakow, master clockmakers had to make clocks that indicated at least Italian and common hours. Clocks in Bologna displayed Italian hours. While much of Tycho Brahe's work was at Uraniborg in Denmark, the last period of his life was spent under the patronage of Emperor Rudolf II in Prague. Johannes Kepler also worked in Prague under Rudolf's patronage.

Implicit in Regiomontanus's and Walther's ideas and technology was an interest in observations serving as the basis for mathematical modeling, which included the ability to adapt observations to different timescales. In a letter to Christian Roder, Regiomontanus complained about those interested in astronomy who, rather than observe the sky, sit in their cottages viewing the Alfonsine tables as a gift from heaven.[73] Because of this he doubted, if not outright rejected, the usefulness of astrology—particularly astrology based on the Alfonsine tables, in which Regiomontanus noted errors and described tables derived from Alfonsine tables as suffering from the "disease of their mother."[74] Regiomontanus's and Walther's interest in observation, instruments, and in understanding the mathematics of circles and spheres was about imagining the globe more than it was about determining the alignments of planets at a particular location. At best, astrology was subsidiary to astronomy for these two scientists, and not an endeavor that should shape astronomy.

The significance of Regiomontanus's and Walther's choice is that they adopted a timescale that was not the timescale associated with Nuremberg, nor the most commonly used timescale at the time in Europe, canonical hours. Instead, they chose a timescale that emphasized uniformity throughout the year, and which could be equally and easily applied in any location, thereby providing an example of how timescales for scientific applications need not, and sometimes should not, be the timescales adopted for governmental and regulatory purposes. But it must be kept in mind that by going against astrology and canonical hours, they were making a conscious choice to go against dominant ideas about timekeeping, particularly among those receiving elite patronage. Regiomontanus's choice succeeded not just because of the utility of his work, but because of his ability as a publisher to mass reproduce and distribute, and his ability as a teacher to train Walther to carry out and expand his work. Walther's reproduction of Regiomontanus's choice succeeded because of his commitment to long-term observations and methodological innovation. In other words, it is not enough to develop ideas about timekeeping; those ideas must be culturally reproduced.

What Regiomontanus's publications reproduced was not just astronomy tables, but a system of coordinates to use when adapting the tables to any locations. Time pluralism was not an obstacle for scientific research; it was an obstacle to imagining the globe in a single coordinate system. But such a global image would erase local differences even as it eventually enabled UTC.

Notes

1. Cassini, *Recueil d'observations*, 27.
2. Ptolemy, *Almagest*; Sacrobosco, *Sphere*.
3. Podestà, "I Primi Oriuoli Pubblici in Bologna," 151–56; Rose, *Italian Renaissance of Mathematics*, 44–45; Rubbiani, "L'Orologio del commune di Bologna."
4. Rose, *Italian Renaissance of Mathematics*, 45.
5. North, "Western Calendar"; Moyer, "Aloisius Lilius"; Ziggelaar, "Papal Bull of 1582".
6. Bendefy, "Regiomontanus und Ungarn," 247.
7. Hayton, "Expertise ex Stellis," 43.
8. Curtze, "Der Briefwechsel Regiomontan's," 327.
9. Stromer, "Hec Opera Fient," 283.
10. Staudacher, *Jost Bürgi*, 70.
11. Stromer, "Hec Opera Fient"; Werner, "Nürnbergs Erzeugung."
12. Biagioli, "Galileo's System of Patronage," 18–25.
13. King, "Astrolabe Presented by Regiomontanus"; *Astrolabes and Angels*.
14. Ramus, *Scholarum mathematicarum*, 65; Zinner, *Regiomontanus*, 135.
15. See Shapin and Schaffer, *Leviathan*, 25.
16. Beaver, "Bernard Walther."
17. See Bilfinger, *Der Bürgerliche Tag*.
18. Ptolemy, *Almagest*, H258–59.
19. Ptolemy, *Almagest*, H142–43.
20. Isidore of Seville, *Etymologarium*, 215.
21. Bede, *Reckoning of Time*, 22–23.
22. Alcuin, "de Disputatio Puerorum," 1,114.
23. Alsted, *Encyclopaedia*, 1,032.
24. Rossi, *Philosophy, Technology and the Arts*, 128.
25. See for instance, Hampson, *Medii Aevi Kalendarium*, 65; Jäger, "Einleitung," 298; Viani, *Elementi di cronologia*, 7; Wöckel, *Populäre Vorlesungen*, 321; Zallinger zum Thurn, *Institutionum juris naturalis*, 3–4.
26. Ammann, *Vergleichung nachuolgender Siben Uhren*; R. F., *Kalendar*.
27. Turner, *Early Scientific Instruments*, 39–45.
28. Melancthon, "Preface," 2.

29. See Bilfinger, *Mittelalterlichen Horen*, chapter 6; Gaab, "Die grosse Nürnbergische Uhr."
30. Keyssler, *Travels through Germany*, 178.
31. Gaab, "Die grosse Nürnbergische Uhr," 27.
32. Gaab, "Die grosse Nürnbergische Uhr," 45.
33. Smith, "Cittie of Noremberg," 215.
34. Gaab, "Die grosse Nürnbergische Uhr," 61.
35. Mummenhoff, "Lutz Steinlingers Baumesterbuch," 61; Baader, "Wann man vor 400 Jahren," 230.
36. Mummenhoff, "Lutz Steinlingers Baumesterbuch," 61.
37. Dohrn-van Rossum, "Uhren, Glocken und Zeitorganisation," 68.
38. Gaab, "Die grosse Nürnbergische Uhr," 69–71.
39. Regiomontanus, *Kalendarium*, 53–55.
40. Doppelmayr, *Historiche Nachricht*, 25, note u.
41. Golinski, *Making Natural Knowledge*, 36.
42. Wallis and Robinson, *Cartographical Innovations*, 3.
43. Peuerbach, *Theoricae novae planetarum*.
44. Alfonso, *Tabule Astronomice*; "*Libro de las tablas alfonsies.*"
45. Chabás and Goldstein, "Introduction"; *Survey of European Astronomical Tables*, 5–11.
46. Heilbron, *Sun in the Church*, 6.
47. Curtze, "Der Briefwechsel Regiomontan's," 324–29.
48. Pederson, *Survey of the Almagest*, 125.
49. Regiomontanus, "Observationes."
50. Dreyer, "Prolegomena," XXII.
51. Turner, *Early Scientific Instruments*, 58.
52. Regiomontanus, *Kalendarium*, 58.
53. Schöner, *Opusculum Astrologicum*, chapter XVIII.
54. Schöner, *Globi stelliferi*, chapter XIIII.
55. Regiomontanus, *Ephemerides*, 2.
56. Gingerich, *Eye of Heaven*, 164 and 200.
57. Curtze, "Der Briefwechsel Regiomontan's," 193.
58. Curtze, "Der Briefwechsel Regiomontan's," 193.
59. Curtze, "Der Briefwechsel Regiomontan's," 196.
60. Rose, *Italian Renaissance of Mathematics*, 30–31; Grössing, *Regiomontanus und Italien*, 1980, 231.
61. Goodman, *Power and Penury*, 50–53.
62. Regiomontanus, *Ephemerides.*
63. Hamann, "Johannes Regiomonatnus," 42–43.
64. Regiomontanus, *Kalendarium.*
65. Lindgren, "Die Bedeutung"; Zinner, *Regiomontanus*, 119–25; Ziegler, *Regiomontanus*, 99.
66. Werner, "Nürnbergs Erzeugung"; Ziegler, *Regiomontanus.*
67. Diffie and Winius, *Foundations of the Portuguese Empire*, 157.
68. Maury, "Martin Behaim's Globe," 437.
69. Davies, "Behaim, Martellus and Columbus."
70. Zinner, *Regiomontanus*, 119-125.
71. Harley, "Rereading the Maps," 528–29.
72. See Anderson, *Imagined Communities.*
73. Curtze, "Der Briefwechsel Regiomontan's," 326.
74. Regiomontanus, "Joannes de regiomonte magistro Christiano," 189.

Chapter 5

Humbuggery and Time Pluralism

Nineteenth-Century Time Gambits

The New Beer Bill

> First Tippler.— "Well, Tom, what d'ye think o' the New Bill?;" Second Ditto.— "Oh, it's not so bad; a man can easily get drunk before ten o'clock, if he makes good use of his time!"[1]
>
> —*North Wales Chronicle,* Saturday, 11 November 1854, page 7

Regiomontanus and Walther lived in an era of public time pluralism and diverse ways to reckon time. There were also diverse hour systems: great hours, small hours, Italian hours, Greek hours, Babylonian hours, and temporal/canonical hours. There were diverse technologies used: sundials, clepsydrae, astronomical observation, and early clocks. Such diversity of technology and hour systems allowed them to make choices to serve their goals, but it also required the ability to convert between different systems. By the nineteenth century, clocks with the small clock hours were widespread, alternative timescales were waning in Europe, mean time had come to overshadow apparent solar time, and railroad and telegraphs had opened the possibility of synchronizing clocks over large distances. Trade, navigation, and colonialism all relied on imagining Earth in terms of spatial coordinates linked to time. In short, the idea of an ideal clock was taking hold even as the performance of individual clocks continued to vary. Despite the emergence of the idea of a perfect clock, the differences between clocks allowed for instances of manipulating clock time for political purposes.

Public clocks became an important part of daily life: Alexis McCrossen points out that "timepieces announced and legitimized the active presence of a variety of interests competing for institutional, social, and cultural power,"[2] and she adds that public clocks did not always agree, that there was no institution with the authority to standardize timekeeping, and that every clock seemed to show its own time.[3] The usual story of the history of timekeeping views these movements toward standardization as signs of progress. Such a narrative that narrowly focuses on the emergence of uniformity and standardization in fact focuses on a few experts, not on the messiness of public clocks that did not agree or personal clocks and watches that varied in performance. In fact, the increased reliance on clocks to know the time rather than on observation of solar time fostered disagreement, since timepieces varied whereas the Sun was the same for everyone in a particular location.

There is a widespread assumption that, with the improvement of clocks and the invention of the telegraph, there quickly followed widespread acceptance of clock synchronization. This was not the case. Peter Galison notes that during the late nineteenth century there was a "struggle to coordinate time" in almost any municipality in North American and Europe.[4] Thomas Allen notes the "uncertain reliability" of nineteenth-century American clocks and watches.[5] Cheryl Wells documents many instances of the lack of synchronization of clocks during the Civil War in the United States.[6] This struggle extended to colonies. For instance, in Hobart, Australia, in the 1830s and 1840s, the three church clocks did not agree.[7] In Bombay in the last couple of decades of the nineteenth century, there was resistance to accepting clock time synchronized to Madras's longitude. Since Madras time was the official legal time, but Bombay time was the preferred social time, there was a multiplicity of clock times there.[8] Where there was no locally recognized authoritative time source, people accepted that their clocks and watches disagreed. In her 1891 travel account, *A Thousand Miles Up the Nile*, Amelia Edwards observed "watches soon differ in a hopeless way for want of the church-clock to set them by, and it is always possible that host and guest may be an hour or two apart in their reckoning."[9]

In the 1860s in England there was no agreement on the correct time for a location, and the use of local mean time and Greenwich Mean Time by different clocks in the same city was common. This resulted in published timetables to assist travelers in knowing the correct local time,[10] and the clockmaking firm of Henry Ellis and Son of Exeter creating a directory to indicate which towns used Greenwich Mean Time (GMT) and which used local mean time.[11]

By the nineteenth century in Europe and North America, the form of the standard came to be unquestioned even if local instantiations of the standard were contested. The 24-hour day beginning at midnight became the standard day. Mass-produced clocks and watches encoded this form. While the form was unquestioned, what constituted midnight was not straightforward and consequently could be subject to various claims. Midnight could be twelve clock hours after solar noon; midnight could be determined by the position of the stars revolving around the celestial pole; midnight could be based on observations at one's location; midnight could be based on observations at another location, say, an observatory; midnight could be determined according to mean time, or it could be determined according to apparent solar time. With the adoption of mean time as an average of Earth's rotation and discussions about which meridian should define clock time, midnight was a fluid idea with no set of conventions to anchor it. There was no longer cosmopolitanism about timescales with different ways of counting hours vying for dominance. Instead, there were tactics and stratagems involved in asserting the authority of one clock's time over that of another. Choice of a clock time remained an important feature of political legitimization.[12] Time pluralism had changed from different institutions keeping their own time or different polities having their own hour systems to competing clocks in a context of growing consensus desiring accuracy in timekeeping. But what was an accurate clock? Was it the one that faithfully represented local mean time, or one synchronized to a distant observatory, or one tied to the operation of the local railroad? Time pluralism persisted in new forms and people still had to make choices.

It is not the case that only scientists make choices about timescales. Politicians make choices to further their agendas. Governments use timepieces to create legally recognized time,[13] but shrewdly manipulating the things that tell time gives a legal advantage. As Emily Grabham states, this is because legal time is created,[14] not given. In fact, legal time is a product of the imagination, not of observation. Timescale pluralism is a consequence of the "more than one-ness" of legal time.[15] Legal time contrasts with most people taking clock time for granted. They assume a close connection between material clocks and their imagination of an ideal clock. In cases where there is timescale pluralism, clocks can be readily manipulated to a person's legal benefit or to another person's detriment. What we are dealing with is not widespread cultural knowledge, but a cultural artifact that unscrupulous actors could use to their advantage. This chapter is not about widespread knowledge of clock time, but about particular people who recognize the difference between ordinary clocks and how they imagine the time on an ideal clock in order to take legal advantage of the naivete, if not the ignorance, of others. According to Thomas Allen, when one explores the uses and abuses of time-keeping technology, the homogeneity of timescales "begins to shatter into myriad fragments of heterogeneous, local, and transient temporal cultures."[16]

Drawing on Saussure's classic *Course on General Linguistics* for inspiration, one gets insights into how representations work culturally. He described words as consisting of two elements: a signifier (the word) and the signified (the meaning). He pointed out that the relationship between a signifier and signified is arbitrary;[17] in other words, established by consensus. While not a word, the cultural representation of time is an example of such an arbitrary, and even self-referential, sign.[18]

As an arbitrary sign that appears to be "real," control over the sign conveys immense power. Outside of science fiction, the idea of controlling time is rarely explored, and in what follows, I shall not be discussing time travel, which involves playing with the relationship of before and after, but merely the control over the sign, and advantages gained from this. This power depends on arcane cultural knowledge. Secrets are a source of power,[19] but with arcane knowledge about time, the knowledge is hidden and arcane not because of an effort to protect it, but because most clock users pay no attention to it. This gives an advantage to those who are aware of the creation, definition, and distribution of time units.

Some examples of manipulating time are innocuous and amusing. Such is the case whenever a legislature faces a deadline to pass a law, and somebody moves the hands of the clock to push back the deadline. This duplicity of time allows for the regulations to be officially followed even when violated according to every clock outside of the legislative chamber (as well as every watch on every person in the room). In the second half of the nineteenth century, the United States Senate often turned the clock back in order to create more time for the passage of pressing legislation.[20] *Riddick's Rules of Procedure*—a guide to parliamentary procedure used by many legislatures—discusses this practice.[21] This guide states that, in situations where motions need to be passed by a particular date and time, some legislative bodies stop their official clock "by agreement of the 'powers that be' without any motion or announcement one minute before the designated hour" but it then adds, "Courts have decreed that action taken after the designated true hour is void."[22]

A notorious example of this was in 1964, when Denmark Groover hung over the balustrade in the Georgia State House to stop the clock in order to allow the legislative session to continue to debate a redistricting bill. When Groover awkwardly hung over the balustrade to stop the clock, his plan went awry: "When I reached for the little knob to turn it off, it fell. It dangled on the cord for a moment, and I tried to hold it, but it slipped." Representative Groover did not fall, but the clock did, and the bill passed despite Groover's opposition.[23] After the incident, Representative Groover complained that the administration, which supported the bill, had already stopped the clock once to allow continuing debate, and that he was merely attempting the same tactic in order to allow one of his colleagues to speak against the bill. He also agreed to pay for a replacement for the clock.

While this story is amusing, the manipulation of the clock for political purposes can have serious effects. This chapter explores two such cases: the attempt by Fernando Wood to determine delegates to the 1860 Democratic Party convention, and the enforcement of liquor laws in the United States and Great Britain in the last third of the nineteenth century. The nineteenth century in Great Britain and the United States is a pivotal period in the emergence of modern time-keeping practices. Clocks and watches began to be affordable to members of the general public—my own ancestor bought his first watch in the 1860s. A simplistic view of history conflates the imagination of clock time and the time displayed on actual clocks. This leads to viewing the mass distribution of timepieces as a sign of an emerging ideology of uniform time, but because of the varying performance of clocks and watches, the distribution of timepieces combined with the availability of telegraph time services for synchronization by the middle of the century created confusion, not uniformity. The ideology of uniformity associated with the ideal clock did not match the reality of the varying performance and use of clocks. There were no time zones, and as railroad travel became important, railroad lines began to adopt their own time standards that could be considerably different from the locales through which the railroad passed. People in major transportation hubs where multiple railroads connected could find themselves having to keep track of several times, such as reported in the *New York Sun*: "A St. Louis man who professes to carry 'railroad time' must have five watches about him: for trains going out of that city are run according to five different standards of time ...If some enthusiast should attempt to carry the railroad time of the country, he would have to struggle under the weight of fifty-five timepieces; for there are fifty-five standards of time in use by the railroads."[24]

Out of such confusion eventually comes legally recognized standards, but the process is slow. Law lags. According to Grabham, law is the "result of processes of (temporary) stabilization of things and relationships" and "the 'time' or 'times' are provisional upon actions, meanings, and states of matter."[25] Attempts at stabilization built on provisional assumptions are bound to be manipulated by those who recognize the arbitrariness of time as a signifier. In this circumstance, laws and rules in relationship to legally recognized objects, such as clocks, manage "law's ambivalent ontological status."[26]

Railroad clocks fostered the movement from this confusion to legal standards.[27] Railroad clocks coexisted with other public clocks, but the railroad clocks differed from these other clocks in how they determined time. The local clocks began as indicators of local mean time whereas railroad clocks represented a synchronized time across many locales. This translocal time eventually became attractive to travelers and lawmakers. But before translocal times became time standards, there was opportunity for manipulating time based on the time source chosen.

Fernando Wood and the New York Democratic Convention of 1859

At the New York State Democratic Convention of 1859, the proslavery politician Fernando Wood took advantage of confusion over the correct time to usurp control and push through a slate of his own policies. To do this he combined several tactics he had learned during his political career with the novel idea of taking advantage of the difference between clocks set by the railroad, personal timepieces, and the clock in the convention hall. His need to resort to the tactic of manipulating clock time was the result of his weakened political position after years of playing one political faction against another.

Fernando Wood was a notoriously corrupt politician associated with the Tammany Hall political machine in New York City. He served as mayor of New York City and as a member of the United States House of Representatives. He had the characteristics F. G. Bailey associated with skilled political operators—characteristics that included "a readiness to transcend (that is, to defy or ignore or distort) the rules that constrain lesser people in their reasoning in their ethical standards."[28] One of Wood's biographers, Allan Franklin, described Wood as "prepossessing, unprincipled, hypocritical, covetous, and smart."[29] Due to his travel and interests, he was also unusually well informed about time differences and the distribution of time signals.

In 1859, the issue of slavery divided the Democratic Party in New York. In the United States, there were three major positions on the issue: those who advocated for the total abolition of slavery; those who advocated for the regulation of slavery and compromise; and finally, there were those like Fernando Wood, who argued against any restrictions on slavery. Despite being a northern state, because of Fernando Wood, the New York State Democratic Party split in the same way as the national party between regulating slavery and opposing any constraints on slavery.

Fernando Wood was skilled at manipulating all these factions to garner support to win nominations and then elections. He even resorted to violence and coercion to sway political conventions, employing men known as "shoulder-hitters," i.e., men who knew how to throw a punch, to coerce nominating conventions to support Wood. In 1854, using these tactics Wood won the mayoral election with a narrow plurality of the vote. The

Morning Courier and Enquirer, a New York City paper associated with the Whigs, commented on his election:

> It now appears that Mr. Wood is Mayor. With a majority against him, Mr. Wood is Mayor. Supported by none but ignorant foreigners and the most degraded class of Americans, Mr. Wood is Mayor. In spite of the most overwhelming proofs that he is a base degrader, Mr. Wood is Mayor. Contrary to every precedent in the allotment of honor through a municipal history of nearly two hundred years, Mr. Wood is Mayor. His assertion to us that a murder by his own hands could not prevent his election had reason in it; Fernando Wood is Mayor.[30]

Through the combined use of intimidation, patronage, and backroom deals, Wood was reelected mayor in 1856. By 1857, his scheming caused tension with Tammany Hall. Its leaders inducted new members into the organization to undermine Wood. With Wood's weakened position within Tammany Hall, the organization revised the primary system and sought the support of President Buchanan in its battles against Wood. The New York State government then limited Wood's term as mayor and formed a separate police force for New York City from the one controlled by Wood—this eventually led to Wood being arrested by the rival force. By the 1858 elections for mayor, Wood could not win the Democratic Party nomination, having been ousted from Tammany Hall and undermined by the state Democratic Party. In response, he formed his own political club named after its meeting place, Mozart Hall, and turned his attention to ensuring a proslavery Democratic nominee for president of the United States.

While the Democratic Party supported slavery in some form, Wood was critical of the stance of Stephen Douglas, one of the leading candidates for the Democratic Party nomination. Douglas advocated for each state to decide whether it was a free or slave state, but Wood did not want any restrictions on slavery.

Wood realized that if he could gain control of the New York delegation to the national convention, there was a chance that a pro-South, proslavery candidate would be nominated instead of Stephen Douglas. Moreover, he hoped that such a pro-South, proslavery nominee would choose Wood as a running mate.[31]

The odds were stacked against Wood. The New York State Democratic convention was controlled by pro-Stephen Douglas politicians.[32] Wood's challenge was to try to create a situation in which his minority faction could appoint delegates. His goal was to change the rules to favor his selection of delegates, but the rules were that the party convention selected all the delegates. As long as this was the case, Wood's antagonists would have their way.

None of this seems to have much to do with time or the strategic use of clocks, but 1859 predates modern time zones in the United States, but not the use of telegraphs to synchronize clocks along railroad lines. In 1859, the United States was divided into a patchwork of local mean times connected by railroads that maintained their own time standards. Often the largest city a railroad served determined the time for the entire line.[33] Many cities, including Syracuse, where the convention would be held, had clocks that displayed different times. Differences between railroad times and local times were common.[34] In the case of Syracuse, there were two public times: the local mean time, and New York Central Railroad time—a common nineteenth-century form of time pluralism. There was roughly a 10-minute difference between the two, with New York Central Railroad time running ahead of Syracuse mean time. Wood deployed a stratagem of cultivating uncertainty[35] about the correct time in order to usurp the opening of the convention.

It is possible Wood got the idea during his time as mayor. In 1856, the Dudley Observatory in Albany approached Wood with a plan to provide a synchronized time ball service to the city of New York with it costing the city only the annual wages of one person to maintain the apparatus. Wood presented this proposition to the City Council on 12 February.[36] The goal was to synchronize the city's clocks to this reliable time source in order to eliminate the "chronic incongruity among the great clocks" of the city. As a result, the *Times* reported, "Men with engagements, or notes to meet, or checks to deposit against, never know by what standard their default may be determined."[37] By September, the City Council and Chamber of Commerce seemed to have forgotten the proposal.[38]

Wood was also aware of the difference between railroad times and local times. This difference came about after several horrific accidents resulting from the sole use of local times. When trains traveling in opposite directions shared much of the same track, they had to coordinate when to turn onto a side track to avoid a collision. If every location on an east-west axis on a railroad line maintained a different time, such coordination was exceedingly difficult. It was this time difference that led to the terrible collisions that occurred on the Camden and Amboy Railroad that killed four and the Providence and Worcester Railroad that killed fourteen within three days of each other in 1853. The Camden and Amboy collision occurred on a single track that went around a sharp curve with limited visibility. Just before the curve was a turnout at which the New York-bound train was supposed to wait if it arrived around 4:12. The coroner's inquest for the Camden and Amboy accident indicated that a disagreement between watches played a critical role in the wreck.[39] The solution the railroads adopted to avoid such crashes was to use a single time along the entire railroad line and to synchronize the station clocks using a telegraph. This is why Syracuse had two times—one reflecting the railroad, which set its clock to New York City's meridian, and the other reflecting the local mean time.

In Syracuse, the New York Democratic Party's convention was set to take place at Weiting Hall, an opera house in the downtown area of the city, strategically located between the Erie Canal and the new New York Central Railroad terminal. At least through 1858, the New York Central Railroad announced in the *Syracuse Daily Courier* that its trains ran on Albany time, rather than Syracuse time, with Albany time being nine minutes ahead of that of Syracuse.[40] The clocks in the railroad terminal displayed the railroad's time, not Syracuse local time.

Wood hatched a plot to take advantage of there being no official time standard. The convention was to start at noon but did not specify whether New York Central Railroad noon or local Syracuse noon was the correct starting time. This ambiguity was a key piece of his effort to hijack the convention. His delegation included some official state delegates and a fair number of "shoulder-hitters" to help him defend his gambit by force, if necessary. These people included John C. Heenan, nicknamed "The Benicia Boy"—regarded as the best boxer in the world. Wood's group also included Hiram Smith, from Monroe County, who was an officially recognized delegate at the convention. Because of this status, Smith played the important role of nominating Wood's man, Thomas Alvord, as chair of the convention. In an affidavit by John Carpenter, who was among Wood's retinue, he recounts that they were to "assist FERNANDO WOOD in forcibly taking possession of the Convention" and "that we were all to place ourselves under the Benicia Boy, who would be our chief reliance when it came to a fight."[41]

At noon New York Central Railroad time, Wood established his delegates in the convention hall. The delegates from other factions were still arriving, expecting the convention to open at noon Syracuse time. Using the time difference, Wood had Hiram Smith call the convention to order and nominate Thomas Alvord as chair. As more of the official delegates arrived, a violent altercation followed in which Heenan punched the Tammany chairman so hard that he needed medical attention, and Wood's followers kept control by brandishing pistols.[42] Wood's followers then passed nominating rules for delegates to the national convention. These rules included creating a new state committee with Wood as a member, the determination of the Democratic ticket for state offices, and a date for the election of delegates-at-large for the national convention.[43] Wood's faction then adjourned and left the hall, at which time Wood's opposition reentered and convened the official state convention. There was concern that splitting the party would strengthen the Republicans, so this official convention decided to keep some of the upstart convention's statewide ticket, but removed Wood from the state committee, and rejected the election of delegates to the national convention in favor of the state convention determining those delegates.[44] Wood's gambit was only partially successful, and despite his losing control over the nomination of delegates, he appeared with a delegation at the national convention in Charleston anyway, on the grounds that his New York State convention was the official one. This created the awkward situation of New York being represented by two sets of delegates.[45]

Accounts of the New York State convention indicate that there was disagreement over which time should be used. The *New York Times* reported that "Fifteen minutes before the time fixed for the assembling of the Convention, a delegate, in the interest of FERNANDO WOOD, moved that Mr. ALVORD, of Onondaga, be made temporary Chairman, put the question, and declared it carried."[46]

The *Buffalo Daily Courier* reported that Hiram Smith "jumped on the floor" when "the clock struck twelve" in order to make a motion.[47] This report, if factual, indicated that at least one nearby clock—either in the hall or close by—chimed the hour according to New York Central Railroad time rather than local Syracuse time.

The *Brooklyn Evening Star* described the event as "The Wood faction, with their notorious leader at their head, obtained possession of the Hall, and jump upon the stroke of twelve, moved one of their own men as temporary chairman and immediately declared him elected in spite of all opposition."[48]

The *New York Herald*'s coverage displayed confusion over the time the convention started. This newspaper was aligned with the Democratic Party, so the confusion in the reporting possibly reflected the divided state of the party. One report reads: "Several minutes before twelve, no one being in Weiting Hall except the Mozart Hall delegation [Wood's faction], one of them moved that Thomas G. Alvord, of Syracuse, be chairman ...At twelve, Peter Cagger, as chairman of the State Committee, called the meeting to order, and moved that John Stryker be chairman."[49] A different report on the same page states, "Immediately on the stroke of twelve, HIRAM C. SMITH, of Monroe, jumped on the floor, and moved that Thomas G. Alvord be temporary Chairman of the Convention ...The main portion of the Convention, who stated it now wanted fifteen minutes to twelve o'clock, here entered the hall." After this, there ensued an argument between the two factions. The account continues by saying, "At this instant PETER CAGGER appeared on the platform with his watch in hand and said: — GENTLEMEN OF THE CONVENTION—The time having arrived for the organization of the Convention, I move that John Stryker, of Oneida, be temporary chairman." In the last column of the page is a report on

Fernando Wood's view of the situation: "At twelve o'clock precisely, by Syracuse time [he was lying since he had used New York Central time], the Convention was organized by Mr. Smith, of Monroe, whose seat was not contested, and who had a right to move the organization …Fifteen minutes after the hour at which the Central Committee had ordered the Convention to assemble, Peter Cagger, not a member of the Convention, had called the Convention to order."[50]

As the *Buffalo Commercial* reported, Wood then took the floor and gave a speech lamenting the violence. In the speech, he observed that the convention "had proceeded to temporary organization in a regular manner at 12 o'clock precisely, by Syracuse time" and that "The Convention was organized by Mr. Smith, Monroe, whose seat was not contested, and how he had a right to move an organization."[51]

By 22 September, the *Buffalo Daily Courier* commented on the issue of time in a small report about a meeting of Republicans in Rome, New York:

> The Republicans of Rome, in their call to elect delegates to the County convention, say it will meet at 7 P. M., "Central Railroad Time." They evidently do not intend to lose the organization by any dispute as to exact difference between New York and Rome time. Wood claims that he called the Syracuse Convention to order at 12 P.M., New York Central Time, while Mr. Cagger went by Syracuse time. Hereafter, to prevent mistakes, the standard of time should be stated in all calls for political meetings.[52]

At the 1859 Democratic State convention, there was a dispute over which time was the official time. The Mozart Hall delegation used railroad time, whereas the Tammany Hall delegation used some other time which featured Peter Cagger waving his watch. The accounts suggest a fifteen-minute difference. This is a bit of a puzzle since the longitudinal difference between Syracuse and New York is only about 2 degrees—a difference that would produce an eight-minute, not a fifteen-minute, difference in local time.

The *New York Daily Tribune* gives an account that enlightens a bit more about the time difference: "the alert Mr. Cagger was outwitted in the preliminary organization by an agile Hard [a political faction] from Rochester, one Smith by name. Peter, struck dumb by the audacity of his rival, at length recovered his senses, and eyeing first the placid face of his watch, and then the perturbed visage of Mr. Dean Richmond, seemed to inquire of each whether it was in order to start the Convention of 'railroad time.' When the hour of 12, local time, arrived, Cagger placed Mr. John Stryker on the platform as temporary Chairman."[53]

While this account suggests that the difference in time was not between two local times but between railroad time and local time, it introduces yet another element of confusion when compared with the first accounts.

According to one account, Mr. Smith organized the convention at 12:00 Syracuse time, and according to another account, it was Mr. Cagger who did so. In another account, Hiram Smith took control of the convention at noon New York Central Railroad time, and Cagger took the stage later based on Syracuse time. Wood's representation of Cagger taking the podium fifteen minutes late might be an exaggeration, or might have been based on Hiram Smith's watch, if Smith set his watch to the time in his home county of Monroe, which was around 1.5 degrees of longitude west of Syracuse, and about fourteen minutes behind New York City time.

Whatever the case, the existence of multiple times was the basis of a tactic to attempt to usurp control over a political convention. Wood made sure his forces were in place before the other factions could muster their supporters. This shows that not only was Wood aware of the different timescales, but also that there was an awareness that confusion over the timescales could be used strategically.

Wood's gambit with the ambiguity over the meaning of *noon* was partially successful. While he did not get everything he wanted, for somebody leading a faction that had no official seat at the table, he gained quite a bit. When the Democratic convention met in Charleston, South Carolina, in 1860, Wood was there with a delegation supporting the proslavery candidate, John Breckinridge. Despite Wood's efforts, the convention deadlocked in choosing a nominee for president. This resulted in a second convention being held in Baltimore, at which Stephen Douglas obtained the Democratic nomination.

Liquor Laws and Time

The choice of clock time is a useful tactic in achieving political goals. David Rooney and James Nye document this with regard to nineteenth-century British liquor laws where morality, the regulation of alcohol consumption, and the use of clocks to shape behavior combined.[54] The organized temperance movement originated in the United States with religious groups but quickly spread to Great Britain. The temperance movement in Great Britain was diverse and included interests from those advocating total prohibition, to those who urged moderation in alcohol consumption and saw a government interest in creating laws to prevent public drunkenness.[55] The efforts of this movement eventually resulted in the Liquor Licensing Act of 1872, which regulated the hours of operation and licensing of public houses but fell far short of the goals of temperance advocates.

This law created debate about time. Closing time in London was at midnight on Saturdays and 12:30 AM other days, but in other places the mandated closing time was either 10:00 or 11:00 PM. Before this law, restrictions on hours varied depending on the municipality, with London establishments open almost all the time. In 1839, the Metropolitan Police Act allowed the city to shut down drinking establishments on Saturday nights, but there were no restrictions on the rest of the week.[56]

The focus on hours of operation in the enforcement of liquor laws led to several cases concerning whether legal time was local mean time or Greenwich Mean Time. By 1852, Greenwich Observatory was sending out telegraph time signals to railroads and the national post office system. The widespread availability of Greenwich Mean Time did not make it a legal time—it only made it the most widely distributed time from a reliable time source. The difference between local mean time and the time provided by a time signal from an observatory might only be a few minutes, but those minutes were enough to find a bar or public house in violation of the law.

By the time of the 1872 Liquor Act, the court case Curtis v. March discussed in the introduction set the precedent for defining legal time in Great Britain, but this was not general cultural knowledge, or even widespread knowledge in the legal community. Curtis v. March was an appeal brought by a defendant who had shown up for his court case only to find that he had been convicted without an opportunity to mount a defense. He appealed the conviction on the grounds that he appeared on time according to the town clock and the mean time of Dorchester, whereas the court's clocks were set to Greenwich Mean Time, which was about ten minutes ahead of local Dorchester time. The court noted that such differences created confusion with many legal ramifications. The judges hearing the appeal determined that the legal time was the time of the location, not Greenwich time.[57] As a result of this precedent, in applying the 1872 Licensing Act, the local meridian and not telegraph time services using Greenwich Mean Time determined the time of day.

Confusion between Greenwich time and local time brought about a cascade in time tactics on the part of sellers of liquor, consumers of liquor, law enforcement, and temperance activists. The time reported on a clock would be a matter of key evidence that would make the difference between a business suffering a heavy fine or not. At the 1873 meeting of the United Towns' Association of Licensed Victuallers in Birmingham, the secretary, H. C. Edwards, reflected on the effects of the Licensing Act of 1872. He said to his audience that "a deliberate design had been formed to first entrap and then ruin." To him, the law was not about the regulation of the sale of alcohol, but instead about the destruction of the businesses that sold it.[58]

There was a real financial threat to those selling alcohol, in part because under the law, municipalities could turn over the money collected in fines to the police superannuation fund. In effect, the more fines the police could impose, the more money flowed into their retirement fund.

To make matters worse for publicans, in addition to the difference between local mean time and GMT, there was not widespread agreement about clock time. In December 1872, there was a case held before the Keynsham Petty Session Court against Albert Sincox, an attendee at a dinner of the 7th Somerset Rifle Corps held at the Lamb and Lark Inn. The event ended at 11:00 PM with the band playing the national anthem, but Mr. Sincox was among those who tarried at the inn, putting him in jeopardy of being charged with being on the premises during prohibited hours. The officer making the charge, Constable Kingcott, stated that he had not seen Mr. Sincox leave by the time the band left at 11:30. The officer was sure that it was 11:30 because he looked at his watch and heard the chimes, but he also testified that the church clock's time ran behind his watch's time. This led to an examination of the poor police officer focusing on his confusion over time.

> Cross-examined by Mr Norris: He knew the time because he looked at his watch and heard the chimes. The chimes went after his watch.
>
> Then your watch was fast? —No; the church clock was not fast enough. (Laughter.)
>
> Your watch is always right, I suppose? –It is Greenwich time.[59]

Eventually, it was determined that the poor constable's watch was actually ten minutes fast, even though he claimed to set it to the clock at the railroad station every day. His choice of setting it to a railway station clock might have explained his watch being fast, however, since it was the practice of some stations in the United Kingdom to have their clocks set ten minutes fast to encourage passengers to get to the station on time for their trains. This practice continues to this day at the Balmoral Hotel near the train station in Edinburgh.

Within London, many public houses had access to time signals from the Standard Time Company and set their clocks accordingly. But London, or for that matter any town near the meridian of Greenwich, did not experience a significant difference between local time and GMT. There were cases of publicans seeking to trick customers by means of setting clocks to run fast, however. One such pub, the Duke of York on Salmon Lane, found itself in trouble as a result. A constable had entered the pub at 1:20 AM on the pub's clock to find people drinking at the bar. The constable said that he knew the time because he had heard Big Ben chime the quarter hour, and that he believed that the clock in the bar was too slow. The lawyer for the defense argued that the constable was mistaken—that the pub's clock was ten minutes fast in order to encourage customers to leave before closing time. Moreover, that the pub was too far from Big Ben (about four miles) to hear the small quarter chiming bells. Because of the confusion over the time, the case was dismissed.[60]

The tactics deployed against publicans involved using time to entrap them. On the very first day of the Licensing Act of 1872, the *Western Daily Press* of Bristol printed a story about two Good Templars (one of several temperance associations) who watched the Three Crowns in Bridgewater to report to the police that there were people on the premises after closing.[61]

By 1873, the confusion over which time was implied by the law became apparent. In a case concerning the Rose and Crown Inn of Ashburton for western England, James Westaway, the owner of the inn, appealed his fine on the grounds that there was a question as to whether "publicans were bound to close by mean time, which was a quarter of an hour before Greenwich, or railway time." In this particular case, the court found the defendant served beer twelve minutes after 11:00 local mean time, but that was three minutes before 11:00 according to Greenwich Mean Time. The witness against the Rose and Crown did not appear, resulting in the case being dismissed.[62] The precedent of *Curtis v. March* would have clarified this case if it had been applied, but the magistrate did not refer to this case.

Then in Crewe in 1874, a town about 2.5 degrees west of London, Thomas Blayley of the Wolverton Arms public house was charged with selling liquor during prohibited hours. The defendant's solicitor asked the police officer if the time he used was that of Greenwich, and the officer said it was. He then put forth the defense that while Greenwich time was used throughout England, the difference between Greenwich time and the mean time of Crewe was about a quarter of an hour (it is actually about ten minutes), and without any reference to *Curtis v. March*, he argued that the law applied to the mean time of the place of the office, not mean time according to Greenwich. At least two different newspapers reported on this case.[63] One of these newspapers reported a similar case in 1877, resulting in the proprietor of the Prince of Denmark beerhouse getting the case dismissed.[64]

By 1878, the precedent of *Curtis v. March* was explicitly applied. In Wigan, the landlord of the Three Crowns was charged with serving alcoholic drinks outside of the permitted hours. His attorney, Mr. Rowbottom, argued that according to *Curtis v. March,* the legal time was the mean time of the location of the Three Crowns, not that of the parish church clock "which had many sins to answer for, as it was very irregular in its actions." The court decided not to fine the landlord under the Licensing Act provisions, but to require him to pay for the court costs.[65]

Law enforcement would catch on to this ruse and use it for its own benefit in enforcing the laws. In Wellington, John Fieldhouse, the landlord of the Hand and Heart beerhouse was convicted. His defense was that he closed his establishment according to his watch, which he set by the church clock, but the prosecution pointed out that the church clock was set to Greenwich time and that, citing *Curtis v. March*, the legal time was the local mean time.[66]

In an article published in May 1880 from York, the Chief of Police stated that the legal time for York was the local mean time, not Greenwich time. He then noted that no clock in the city displayed local time—by 1880, they all had adopted Greenwich Mean Time.[67] This suggested that the police knew the legal time even if the drinking populace did not.

In mid-June 1880, Dr. Charles Cameron, MP from Glasgow, introduced a bill to Parliament to establish Greenwich Mean Time as the legal time of the United Kingdom. Cameron had been associated with the temperance movement in Scotland. Shortly after being elected to Parliament in 1874, he was a cosponsor of a prohibition bill that eventually failed. In 1878, he also was the primary sponsor of the "Habitual Drunkards Bill," which supported treatment for alcoholics—Parliament passed this bill. The confusion over the legal time had created tactics by which those who sold alcohol could defend themselves. Cameron sought to eliminate that confusion by defining a national legal time based on Greenwich. Cameron sponsored the "Statutes (Definition of Time) Bill," which passed the House of Commons with little debate on 8 July 1880, and then became law on 2 August. In fact, an argument could be made, and has been made by David Rooney, that the pressure to adopt standardized time zones did not just emerge from the railroads, but from liquor laws, "because anti-alcohol reformers wanted to use clocks to police their moral crusade."[68]

Conclusion

These are not examples of widespread cultural knowledge or cultural consensus, but of time pluralism causing confusion over what the correct time was—confusion cultivated by a growing acceptance of the imagination of the ideal clock even as material clocks varied in the time they represented. For many, time is naturalized, and the time on the clock is not questioned, but for some—particularly those who understand clock time as a created sign—the time on the clock is something that can be manipulated. Where there is time pluralism, there is opportunity for manipulation. Fernando Wood, astute publicans, informed temperance activists, and clever police all made use of time pluralism to achieve their goals. It is easy to look upon this period and say that everything was resolved by virtue of laws passed defining legal time, but to do so is again to demonstrate how easy it is to not be aware of the legal standing of time or how time pluralism can hide behind devices to which people attribute naturalized time. For instance, UTC has only been the legal time in the United States since 2007. Granted, this pluralism is at the level of fractions of a second, but with a financial sector in which immense amounts of money can be acquired or lost in a millisecond, the difference of a second looms large.

Notes

1. *North Wales Chronicle*, 7.
2. McCrossen, "Conventions of Simultaneity," 219.
3. McCrossen, "Conventions of Simultaneity," 10, 17, 42, 44.
4. Galison, *Einstein's Clocks*, 40.
5. Allen, *A Republic in Time*, 98.
6. Wells, *Civil War Time*.
7. Davison, "Punctuality and Progress," 178.
8. Mehta, "Standard Time vs. Bombay Time"; Masselos, "Bombay Time."
9. Edwards, *A Thousand Miles*, 89.
10. Esbester, "Designing Time."
11. Thrift, "Capitalist Time Consciousness," 125.
12. Lazar, *Out of Joint*.
13. Grabham, *Brewing Legal Times*, 6
14. Grabham, *Brewing Legal Times*, 6.
15. Grabham, *Brewing Legal Times*, 23.
16. Allen, *Republic in Time*, 10.
17. Saussure, *Course in General Linguistics*, 113.
18. Birth, *Time Blind*, 69.
19. Luhrmann, "The Magic of Secrecy."
20. McCrossen, *Marking Modern Times*, 49–51.
21. Riddick and Butcher, *Riddick's Rules*, 186–87.
22. Riddick and Butcher, *Riddick's Rules*, 187.
23. *Atlanta Constitution*, 1.
24. *New York Sun*, "Only 8 Watches instead of 55," 8.
25. Grabham, *Brewing Legal Times*, 34.
26. Grabham, *Brewing Legal Times*, 44.
27. Gordon, "Railway Clocks."
28. Bailey, *Humbuggery and Manipulation*, 7.
29. Franklin, *Trail of the Tiger*, 43.
30. quoted in Connable and Silberfarb, *Tigers of Tammany*, 122.
31. Mushkat, *Fernando Wood*, 90.
32. Mushkat, *Fernando Wood*, 90.
33. Stephens, *On Time*, 103.

34. McCrossen, "Conventions of Simultaneity," 94.
35. See Bailey, *Humbuggery and Manipulation*, 129.
36. *New York Herald*, "Municipal Affairs," 1; *New York Times*, "Message from the Mayor," 8.
37. *New York Times*, "Standard of Time," 4.
38. *New York Times*, "Astronomical Time," 4.
39. *New York Herald*, "Collision," 1.
40. For instance, see *Syracuse Daily Courier*, "New York Central Railroad," 4.
41. *New York Time*, "The Row at Syracuse," 5.
42. Bonham, "Election of 1860"; Alexander, *Political History*, 257–58; Mushkat, *Fernando* Wood, 89–91; Johnson and Smith, *Political and Governmental History*, 434.
43. Mushkat, *Fernando Wood*, 91.
44. McGuire, *The Democratic Party*, 341.
45. Democratic Party, *Official Proceedings*, 6.
46. *New York Times*, "Proceedings of the Convention," 1.
47. *Buffalo Daily Courier*, "Latest News By Telegraph," 3.
48. *Brooklyn Evening Star*, "The Harmonious Democracy," 2.
49. *New York Herald*, "What Was Done in the Convention," 3.
50. *New York Herald*, "Details of the Proceedings of the Convention," 3.
51. *Buffalo Commercial*, "Democratic State Convention," 1.
52. *Buffalo Daily Courier*, "Fast Time," 2.
53. *New York Daily Tribune*, "Democracy at Syracuse," 4.
54. Rooney and Nye, "'Greenwich Observatory Time'"; Rooney, *About Time*, 148–57.
55. See Harrison, *Drink and the Victorians*, particularly chapter 6.
56. Harrison, *Drink and the Victorians*, 58–59.
57. Hurlstone and Norman, *The Exchequer Reports*, 719–20.
58. *Birmingham Daily Post*, "United Towns' Association," 6.
59. *Western Daily Press*, "New Licensing Act," 3.
60. *Morning Advertiser*, "Dismissal of Two Summonses," 6.
61. *Western Daily Press*, "New Licensing Act," 3.
62. *Western Times*, "Appeal Court," 3.
63. *Birmingham Daily Post*, "Important to Publicans," 5; *Wrexham Advertiser*, "Important to Publicans," 3.
64. *Birmingham Daily Post*, "Aston Petty Sessions," 6.
65. *Manchester Evening News*, "Publicans and Closing Time," 4.
66. *Wellington Journal*, "Selling Ale," 6.
67. *York Herald*, "What is Local Time," 4.
68. Rooney, *About Time*, 151.

Chapter 6

Time Pluralism and Current Timescales

Timescales embrace the beauty of the timekeeping art.[1]
—Parameswar Banerjee and Demetrios Matsakis,
Introduction to Modern Timekeeping

If time-keeping tools tell stories, what stories do our current timekeeping tools and techniques tell and what do these stories conceal? The ideal of clock time is unnatural. After all, as the sociologist Robert Levine observed, "life on clock time is clearly out of line with virtually all of recorded history,"[2] and some consequences of uniform, continuous clock time, such as the 24/7 society, are psychologically and physiologically harmful.[3] Yet, the idea of clock time has become so naturalized that there is little reflection on why it is the way it is, and confusion over its relationship to actual clocks and cycles in the world. Indeed, many think that the uniformity and continuity of the ideal clock represent Earth's rotation even though that tie is a weak one that many time metrologists are seeking to sever (see chapter 8). The power of clocks over our lives contributes to the view of clock time as not arbitrary. Questioning that time reveals the extent to which it is a product of human choice, and not merely a product of observations about the world around us.

To use a clock to know the time is a mundane act. Where does that time come from? *How is that time produced?* The time on the clock is, indeed, produced and not observed. It is not something "out there" that time metrology measures but is the result of a coordinated effort of many national time laboratories distributed across the globe. Many of these laboratories manage more than one atomic clock. If one looks at clock time as the result of production and then explores how it is produced, what one finds looks very unclocklike. When I saw my first rubidium fountain, what I saw was a tall metal container that reminded me of a whiskey still. Hydrogen masers are literal black boxes. Cesium beam clocks are tubes mounted in a frame connected to various gizmos. Such are the naive, first impressions of the equipment that is used to produce UTC. While all these things measure time with impressive precision, none of them display the time standard. They are sophisticated devices that count atomic transitions. Suites of these devices provide output which is averaged and then converted into a local form of UTC. This local UTC is then communicated to the International Bureau of Weights and Measures (BIPM) be means of a time transfer protocol. It might seem surprising, but the BIPM houses no clocks that determine UTC.[4] Instead, the BIPM calculates a weighted average based on reports from time laboratories, each a local representation of UTC, and then lets the world know the error in each time laboratory's version of UTC through the online publication called *Circular T.* As the former Director of Time of the International Bureau of Weights and Measures, E. Felicitas Arias, put it: "*Constructing* a time scale is a bit like assembling a puzzle where the various pieces have to match to render it stable, reliable, accurate, continuous, and accessible."[5]

The ideal clock's timescale reflects a preference for uniform units of time. This preference influences the design of clocks and the definition of timescales. What links actual clocks. the imagined ideal clock, and UTC is the field of time metrology. This is not merely the science of measuring time, but, as Latour notes about metrology in general, it is also a form of universal agreement "*Provided* you find a way to hook up your local instrument to one of the many metrological chains whose material network can be fully described."[6] This metrological chain links individual clocks to a set of practices, not to a single clock to which all other clocks are set. These practices seek to match the output of physical clocks with the concept of the time associated with an ideal clock. This chapter explores the currently dominant timescale, Coordinated Universal Time (UTC), in terms of how the ideas of clock time come to influence people through their clocks and how this timescale came to be agreed upon.

The Functions of Timescales

Without timescales, calendars and schedules would be difficult to standardize, and the standardization of time-keeping techniques is important for the standardization of human social behavior.[7] Moreover, the shared time synchronization of our current experience depends on such standardization. Without timescales, it would be difficult to construct chronologies to indicate simultaneity and the sequence of events. It is these techniques that create the impression of a global present.[8]

UTC is a global reference standard timescale. A reference standard has several features. First, it must be a constant. Second, in the case of timekeeping, it is the result of a standardized means of determination—the measurement of the frequency of cesium-133 in a vacuum. Finally, it must be agreed to by convention as the standard to which practical efforts at measurement refer. Through material clocks that represent UTC, UTC is part of almost every activity that uses electronics. Electronics work using timestamps to record actions and sequence tasks; many also use frequency measurements tied to the standard second. All such technology relies on UTC to function. Power grids, financial markets, information technology, and telecommunications all use the UTC standard. Everyday electronics such as a smart refrigerator, a tablet computer, a car with GPS—basically the vast majority of electronic gadgets—involve conformity to the uniformity and continuity of the ideal clock by means of network systems that synchronize to UTC.

This goes back to the very beginnings of institutionalized time metrology in the nineteenth century. During those early stages, Charles Sanders Peirce pointed out that instruments, rather than direct observation, reveal much of what is known about the world.[9] Peirce was a pioneer in defining metrological units in relationship to a universal standard as opposed to an object—he did this by relating the meter to the wavelength of a spectral line.[10] A universal standard is a goal of metrology—a means of determining a unit of measure through procedures that can be applied anywhere in the universe rather than relying on a specific artifact, such as the now-obsolete meter bar in Paris that used to be the standard measure of length. Consistent with Peirce's view that instruments mediated observation, he developed a pendulum to measure the intensity of gravity and magnetism.[11] Peirce's later semiotics were shaped by his geodesic work with instruments as signs that mediate between abstract ideas and a material reality. The consequence of Peirce's logic is that there is no *thing* that is the standard. Measurement tools and protocols *produce* the standard, but they are never the standard themselves.

If timescales are cultural constructions and dependent on technology and protocols, then they are products of social institutions that meld technology to social relations and not of the ability of any one human being to observe natural phenomena. Currently timescale must have the following features: 1) a starting point ratified by the World Radiocommunication Conference (WRC) of ITU-R, 2) uniform units of time approved by the General Conference on Weights and Measures (CGPM), and 3) a sequence of units that extends from the beginning point into the future or to a point where the timescale repeats that can be represented in computer code with a unique code for every increment of time extending infinitely into the future and documented in the standards created by the Institute of Electrical and Electronics Engineers (IEEE).

Controlling time is a vital element in controlling power—an idea manifest in the close relationship of imperial rule and the definition of time, whether it be Julius Caesar and his calendar reform or the British Empire and GMT. Information technology shapes how electronic devices influence their users' behavior. The standards that allow computers to synchronize in order to exchange data leads to network time—what Robert Hassan describes as a relationship with timekeeping that depends more and more on how network-mediated times influence digital connections between people across space.[12] Network time steers electronic clocks to coincide with the idea of continuity and uniformity associated with the ideal clock. This time becomes "an unreflected-upon clock as backdrop to modern life."[13] Policy decisions about timescales and their distribution subtly influence anyone who relies on an electronic device for the time. As such, a timescale that links one's clock to the ideal clock is a form of what Michel Foucault calls *capillary power*—power that "reaches into the very grain of individuals, touches their bodies and inserts itself into their actions and attitudes, their discourses, learning processes and everyday lives,"[14] and it manifests political decisions that employ scientific knowledge so that information technology functions smoothly.

The timescale of an ideal clock relies on a technological infrastructure. In the case of UTC, time laboratories compose a crucial element of this infrastructure. The distribution of these laboratories across the globe is not uniform so that a few nations house the most significant contributors to the calculation of UTC. As a result, the northern hemisphere plays a disproportionate role in defining time for the portion of the universe occupied by humans and their technology. Synchronized time is a privilege of sorts, even as it is a form of capillary power over one's actions.

The common means of accessing the UTC timescale is through some form of information technology. Standards for code in such technology are rigid—they are "more stringent than rules (laws and technical standards) written on paper."[15] If one violates a law, regulation, or rule, one only gets punished if discovered. Computers do not allow violations of code—such violations receive an almost immediate error message if they do not lead to more catastrophic results, like system failures. Because current electronics rely on standards for programs and devices to interface, such standards form a very subtle, inviolable, and powerful means of structuring and disciplining thought, and consequently behavior. They are a stealthy means of making electronic clocks simulacra of the ideal clock.

The Function of Standardization

Visible or audible time signals to which people set their clocks coordinate social organization. Now, increasingly, electronics coordinate social organization by commands executed by mostly hidden time protocols. Rigid and robust cyber-synchronization has replaced most individual human agency in setting time and reckoning time[16]—as Paul Virilio and Sylvere Lotringer observe, "Synchronization is bringing standardization to completion."[17]

Standards are necessary for our electronically coordinated world to work seamlessly,[18] and as a source of a standard, a perfectly performing clock is ubiquitous in imagination. This was not so in the relatively recent past. In reflecting on the world before the Treaty of the Metre in 1875, Peter Galison notes that "Each country's peculiar system of weights and measures made comparison among them impossible without tedious calculation."[19] Current ways of coordinating the different facets of our lives would be difficult without agreed-upon timescales. Moreover, standards have power not due to governments trumpeting their power, but due to people using the standards without reflection in their everyday lives.[20]

The Adoption of Uniform Standards

Clock users that rely on a synchronized and idealized uniform time are relatively new phenomena. The ideal clock assumes global uniformity, and time metrology has been moving consistently toward that ideal for over a century. For instance, before 1880, there were several public clocks in England that displayed both Greenwich time and the local mean time. Tom Tower on Christ Church at Oxford did this,[21] and the clock of the Corn Exchange in Bristol still has this feature. Derek Howse determined that "by 1855 98 per cent of the public clocks in Great Britain were set to GMT,"[22] and as the last chapter made clear, this was without any law mandating a nationwide time zone. In fact, there was no standard time defined for the United Kingdom before 1880. A similar situation existed in the United States, where time zones were only defined in 1883 by the General Time Convention, a meeting of representatives from the railroads. 18 November 1883 was dubbed "the day with two noons" and newspapers around the country reported on the change from the many local times to four time zones—usually noting how smoothly it went. A few papers reported on how complicated the change actually was since, previous to this date, each railroad kept its own time, and each city, as well. The *Daily Evening Bulletin* of Maysville, Kentucky, was particularly detailed in its discussion the day before the change took place:

> That many schedules will have to be changed is due to the fact that many connecting roads now using different standards will, after the 18th, use the same. The following is the list; the letter f denoting that the clock is to be set ahead, and the letter s that it is to be set back:
>
> Atchison, Topeka and Santa Fe, east of Dodge City, clocks only, 9 minutes, f.
> Baltimore and Ohio (west) both clocks and schedules 28 minutes, s.
> Boston, Hoosac Tunnel and Western, both clocks and schedule 4 minutes, s.
> Boston and Albany, clocks only, 16 minutes, s.
> Canadian Pacific, (Eastern division) clocks only, 6 minutes s.
> Central Vermont, both clocks and schedules, 12 minutes, s.
> Chesapeake and Ohio, both clocks and schedules, 8 minutes, f.
> Chicago and Alton, clocks only, 9 minutes, s.
> Chicago and Grand Trunk, both clocks and schedule, 9 minutes, s.
> Cleveland, Columbus, Cincinnati and Indianapolis, both clocks and schedules, 28 minutes, s. [23]

The article goes on to list the adjustments for another nineteen railroads.

This standardization was not a legal one, however. It was an agreement between the railroads, and in the United States by 1883 many people followed the railroads' example.[24] And while there was a close link between railroads' efforts at imposing time standards with legal sanctions,[25] it is too easy to look back on this period in terms of the dominance of railroad time and not the friction between local times and railroad times—something that chapter 5 explored in detail.

The friction between different timescales led to laws that adopted railroad times as the backbone of legal times. For instance, a federal law establishing a legal time standard for the United States only passed in 1918. Between 1883 and 1918, there were numerous court cases in the United States similar to *Curtis v. March* in Great Britain. In these cases, courts debated the legal definition of time, and they wavered between a time zone time and local mean solar time.[26] There was no consistency. In 1889, the decision in the case Henderson v. Reynolds in Georgia determined that the court should follow local solar time, not standard time.[27] In contrast, in 1911, the decision in the case *Salt Lake City v. Robinson* held that for the enforcement of liquor laws, standard time was the legal time, not local solar time.[28] The legal tome *The American Ruling Cases as Determined by the Courts* published in 1912 devoted several pages to discussing what it called "conflicting rules" about whether standard time or the mean solar time of the local meridian was the legal time.[29] As Geoff Gordon points out, the relationship between time and law is now taken for granted and "black-boxed," meaning that the choices behind the relationship of time and law are "marginalized or ignored."[30]

The desire for an ideal clock and its accompanying standardization, then, is not a product of the state, although the state appropriated it. It predates and transcends law and state intervention. There is something about the idea of abstract time that shapes behavior to seek a coordination of clocks that affirms the existence of the ideal clock. There is some aspect of interaction with actual clocks and watches that, when coupled with a metrological mindset, cultivates the desire for this abstract, absolutely reliable time.

Horology versus Time Metrology

The idea of a timescale is a significant concern in time metrology. This field seeks to provide a means for accurately and precisely measuring time, or, as I have been describing it, matching clocks to the idea of the ideal clock. The creation of timescales for various forms of measurement is now part of the standard work of time metrologists. Like the term *timescale*, the field of time metrology seems relatively recent, not because there was no attention paid to time measurement before the nineteenth century, but because, for a long time, time metrology was part of horology—the science of measuring time with clocks. At first glance, it seems that time metrology and horology are one and the same, but they have different emphases. Horology seeks to build better clocks, whereas time metrology seeks to build improved standard measures of time. Horology is about clocks; time metrology is about clocks in relationship to the standards of timekeeping.

Whereas a typical clock buyer considers the accuracy of the clock that is bought, time metrologists do not trust any single clock and rely on statistical analyses of ensembles of atomic clocks to determine the time in relationship to a calculated standard that approaches what I am calling an "ideal clock." In fact, as Bernard Guinot, the former director of the International Bureau of the Hour (BIH) wrote in 1973 about Coordinated Universal Time (UTC), "no real clock runs exactly on UTC."[31] The reason for this is that no single clock produces UTC. Instead, the Time Service of the BIPM uses reports from a large number of time laboratories distributed across the world. The Time Services uses these reports in an algorithm that produces International Atomic Time (TAI), and to TAI is added the number of leap seconds (see next chapter) that have occurred since 1972, the beginning of the present epoch.[32] The resulting timescale is UTC. "UTC itself exists only on paper"[33]—an observation made by the chief scientist of the Time Service of the United States Naval Observatory, and two physicists of the Time and Frequency Division of the United States National Institute of Standards and Technology. The clocks that contribute to UTC are local representations of UTC and are referred to with the abbreviation UTC(k), with k being the name of the clock or suite of clocks. For instance, the USNO calls the version of UTC it generates "UTC(USNO)." The BIPM's publication *Circular T* indicates the difference between each UTC(k) and UTC, but metrologists, engineers, and systems administrators do not use this for their daily time synchronization needs. One way to look at this is that time metrologists compare their clocks to a BIPM calculation that represents what an ideal clock would display on a regular basis, with the performance of the actual clocks in relationship to the calculation of UTC noted and distributed. So instead of the time of an ideal clock being what synchronizes clocks, it is local versions of UTC.[34] Unlike horology, time metrology is not just about improving clocks, but about improving measurement techniques and technology, time calculations, the coordination of clocks, the determination of uncertainty, and local representations of UTC. While improving clocks is part of this, so is improving algorithms and time transfer techniques.

The goal is for reliable clocks to indicate time anywhere in the universe. This is the definition of an intrinsic standard—one that does not rely on a particular artifact.[35] Time measurement is one of the most important standards, however, since metrologists attempt to derive all other standards of measure from some combination of time and the speed of light. Ultimately, time will be the standard on which all other metrological standards are based. Even the kilogram is now defined by the General Conference on Weights and Measures (CGPM) in terms of the Planck constant, a constant in physics that relates mass to frequency, with frequency being a function of time.[36]

Whereas horologists advertise their innovations and how they result in a more reliable watch for consumers, time metrologists advertise the precision they achieve, but not the exact technology behind achieving that precision. In fact, because precise time measurement is critical to the military and national security, some of the cutting-edge technologies of time metrology in the United States are classified. The International Bureau of Weights and Measures can chart how well the clocks of the United States Naval Observatory (USNO) or National Institute of Standards and Technology (NIST) perform but cannot know all the working parts of those clocks.

Thus, horology and time metrology intertwine, but their goals and methods differ. Time metrology relies on statistical representations of time produced by sampling time measurements from a large and robust set of high-quality clocks, while horology provides the public with high-quality clocks and watches.

Maybe it is just coincidence, but it appears that the concern with defining timescales unfolded at the same time as the separation of time metrology from horology. In this separation, metrology became concerned with defining a better protocol for measuring time rather than just building a better clock. It embraced an ideal of the clock, against which actual clocks would be compared. In the attention given to the unit of measure, it eventually focused on the definition of the second. The choice of this unit as the fundamental one for the measurement of time would have significant repercussions in global time policy, as well as ongoing discussions of future time standards.[37]

Instruments That Shape Timescales

Techniques and technologies of time measurement are examples of how the relationship between humans and cognitive tools constitutes knowledge.[38] As Don Ihde points out, in such cases, the instrument becomes a means to perceive the world.[39] It also becomes a part of a larger network of people and technologies[40] that not only shapes perception but forms technoscientific social networks.[41] Successful metrological instruments, such as clocks, are "public, regular, and reliable."[42] While the tendency is to view advances in science as due to developments in theory followed by empirical research, often new scientific insights emerge from the use of new instruments that result in researchers seeing their work in new ways.[43] The use of new technologies sometimes produces discoveries. These discoveries then lead to revisions in how units of measure are defined. As Alfred North Whitehead wrote, "The reason we are on a higher imaginative level [in science] is not because we have a finer imagination, but because we have better instruments."[44] In many ways, it has been instruments that have encouraged the idea of the perfect clock as a calculated (not observed) representation of an abstract and uniform time untethered to any observable natural cycle.

An example of how technological innovation leads to changes in standard units is the shift from the day to the second as the fundamental time unit. Originally, time units on clocks were fractions of a day. As clocks improved, the rotational day became a phenomenon to be measured. This led to improvements in ephemerides—tables that predicted the future position of a celestial object in relationship to time—and these tables included predictions of the rotational day. Coupled with printing, ephemerides became widely distributed. Such tables were useful in celestial navigation,[45] but more importantly, these tables were critical for comparing predictions to observations in order to hone formulas.[46]

The measurements used to create ephemerides posed a problem for the use of the day as the fundamental unit of time. The measurement of celestial phenomena needed astronomical constants,[47] but improvements in measurement technologies and techniques revealed that the length of the rotational solar day is not constant. It varies slightly throughout the year as it moves through space relative to the Sun. This variation led to the use of mean time—a timescale based on the average length of day throughout the year.

Then, in the 1930s, with improvements in the precision of clocks, astronomers discovered that not only did the solar day vary throughout the year, but also so did Earth's rate of rotation over multiple years.[48] Scientists attributed this variability and the then documented slowing of Earth's rotation to tidal braking of Earth's rotational movement. High tide is a bulge in the oceans' water that creates a slight braking effect against the atmosphere. The result is a nonlinear small deceleration of rotation.[49] With increased accuracy and precision of time-keeping technology, the observation of the deceleration of Earth's rotation through about 1970 led to the recognition that timekeeping based on units of uniform duration would diverge from Earth's variable rotation.[50] To make the relationship between Earth rotation and a timescale even more tenuous, in recent years Earth rotation has sped up slightly. This led E. Kenneth Seidelmann and Dennis McCarthy to state: "The recognition of the variability of the rotation of the Earth ...meant that measuring time based on the Earth's rotation using angle measurements of celestial objects would no longer be suitable for those who required the most accurate time scales."[51] Basically, the realization that Earth's rotation varied meant that the day could no longer be regarded as a fundamental unit of measure. To achieve uniformity, the ideal clock had to be freed from Earth's rotational cycles.

Gerald Clemence, who was the astronomer and scientific director of the USNO, proposed a solution. He observed that "the solar and planetary theories give the coordinates of the Sun and planets as functions of Newtonian time, whereas the observed positions are functions of variable time."[52] Newtonian time was Newton's idea of an absolute, uniform time that can be used to measure any variable phenomena.[53] So Clemence advocated the creation of such an absolute timescale.

First, he advocated adopting the definition of the day calculated by Simon Newcomb, a mathematician and astronomer who directed the United States Nautical Almanac Office. In 1895, Newcomb used several centuries of observational data about Earth's rotation to create extensive tables of the Sun, and from these he derived a formula by means of which he calculated the predicted length of day for 0 January 1900 (which is de facto 31 December 1899).[54]

Second, Clemence suggested that since Earth's rotation was not a constant, the standard adopted should be based on a fraction of the tropical year.[55] The tropical year, a concept dating back at least to Ptolemy,[56] is defined as the time between the two moments the Sun passes the celestial equator at the time of the spring equinox in the northern hemisphere. This duration is relatively constant.

Third, since the tropical year is too long a duration to be used for clock time, Clemence advocated dividing that year not into days, but into a representation of the second identical to 1/86,400th of Newcomb's ephemeris day. This standardized second would then constitute the basic building block of a uniform timescale. Observation of the Earth's rotation vis-à-vis the Sun (solar time) or stars (sidereal time) would then involve the determination of the difference between the observed rotational period and the absolute, uniform timescale Clemence proposed based on Newcomb's work. This absolute time came to be known as *ephemeris time*. The NIST physicist Judah Levine wrote of this timescale: "Ephemeris time was completely uniform in principle and was independent of the irregular variation in the rotation of the Earth."[57] The ideal clock and its timescale began to achieve its freedom from Earth's cycles.

The idea of an ideal (and perfect) clock is not merely about clocks, then, but is about how technical innovations that improved time-keeping technology led to insights that then motivated policy decisions and changing definitions of standards. Technological innovation leads to new approaches and knowledge that destabilize standards.[58] The sociologist of standards, Laurent Thévenot, argues that such destabilization then produces "moments of doubt" that instigate a decision-making process.[59]

Technology and Choice in Timescale Definition

The consequence of rejecting Earth's rotation as the source of a time constant, and the eventual choice of using the ephemeris second, was that Earth's current rotational periods did not determine the standard timescale in any direct fashion. This means that there is nothing a person can observe in the environment on a day-to-day basis to know the correct clock time. Clock time is no longer reliant on an individual's or even an expert's observations of Earth's rotational day. Instead, metrologists desired to create a foundation for the ultimate timescale that could hold true anywhere in the universe rather than having to be tethered to our wobbly planet. But, to achieve this, the timescale could not be tethered to Earth's tropical year, either. The standardized second had to be defined in some fashion not relying on the measurement of Earth's movements. Eventually, they defined this second in terms of atomic transitions.

In 1967, the CGPM officially defined the second in terms of transitions of cesium-133 atoms.[60] Around this time, articles appeared discussing the problem of maintaining an atomic timescale alongside a terrestrial timescale. George Hudson, who became the assistant chief of the Radio Standards Physics Division of the National Bureau of Standards (NBS), advocated for time pluralism, writing "It should not be held that UT [universal time] is the only kind of time for 'everyday' use," and that "the invention of physically different time scales, such as a nuclear-based one, should be regarded as a desirable scientific activity; but the proliferation of artificial derived scales is to be deplored."[61] In writing about a discussion related to measuring light by means of atomic timekeeping, Hudson and his colleague William Atkinson, another NBS scientist, noted that two systems for measuring time could exist, as long as they "differ appreciably so that there would be little possibility of forgetting to include a conversion factor" at which point somebody said, "there are in use at present for scientific purposes seven types of time."[62]

One of the inventors of the atomic clock, Louis Essen, saw the matter differently. Astronomical and atomic time do not agree, and in Essen's view, to maintain both as separate timescales would "lead to confusion and a duplication of effort."[63] But there was an additional reason for the disagreement between astronomical and atomic time from the very beginning of the adoption of the atomic standard. When the CGPM adopted an atomic definition of the second, it was already three-quarters of a second per year too short.[64] This was due to the fact that the second defined in relationship to the tropical year was shorter than 1/86,400th of a mean day in the year of its adoption,[65] because the length of day produced by Newcomb's calculations was shorter than the observed length of day in the 1950s when the second was defined in relationship to the tropical year. This length of day was shorter still from when the CGPM defined the second in terms of cesium transitions in 1967. How to reconcile an atomically based timescale and Earth's rotational period became an ongoing source of discussion.

Coordinated Clocks and the UTC Timescale

Despite the resistance to maintaining multiple standard timescales, creating a robust system of time measurement on a rotating geoid with a circumference of almost 25,000 miles requires expert awareness of the variability in performance of material clocks in order to achieve uniformity. In creating a timescale, the comparison of high-precision atomic clocks is crucial.[66] Clock comparison raises several technical challenges stemming from time signals being broadcast to a central location to be compared. One of these challenges is that the Earth is in constant motion, so the signals travel in relationship to a moving Earth. If a signal travels against the rotation of the Earth, the distance it covers is shorter than the geographic distance between the two points. Conversely, if it travels with the rotation of the Earth, the distance covered is greater. Because the origin, target, and signal are all in motion, there is a Doppler effect on the signal itself that needs to be considered when dealing with high levels of precision.[67]

Relativity poses an additional challenge. Since the relative speed of the signal is much greater than the relative rotational speed of its origin and destination, as the theory of relativity predicts, time dilation affects the temporal relationship of the signal and the receiver of that signal. An experiment conducted in 1971 by the astronomer Richard Keating and the physicist Joseph Hafele demonstrated the reality of time dilation. It involved two atomic clocks. They placed one clock on a plane, and the other remained at the airport. The plane then flew around the Earth, and by the time it landed, there was an average 66 ns difference between the two clocks when flying eastward (against the direction of Earth's rotation), and an average 205 ns difference when flying westward (with the direction of Earth's rotation).[68]

With time dilation at the level of nanoseconds occurring with simply the speed difference between an airplane and a point on the planet, with levels of precision now exceeding nanoseconds, and with signals traveling near the speed of light, the amount of time dilation in the signals is much greater than on the planes, creating challenges for the transfer of precise time information. In 1974, Victor Reinhardt of NASA's Goddard Space Flight Center argued that "the precision of global timekeeping is approaching the level where one should consider the relativistic effects of the rotation of the Earth on remote clock synchronization."[69] Fundamental to this challenge is the difference between the time in a particular location, known as *proper time*, and a timescale applied to multiple locations, which is known as *coordinate time*. To create a coordinate timescale, one must consider relativistic effects. This is a facet of clocks versus the ideal clock. Because of current levels of precision, the transfer of time information across the globe and to spacecraft and satellites within the solar system is subject to converting local proper time measurements into coordinate times by means of calculation.[70] Because of the relativistic issues raise by time transfer and the use of proper times from locations around the globe to calculate a coordinate time for the globe, in 1991, the BIPM started to incorporate the theory of relativity into its algorithms.

The Doppler effect and relativity result in incredibly small inaccuracies in time measurement and synchronization, but current time measurement technology allows for recording incredibly small increments of time. As a result, to calculate a reliable timescale requires taking into account these factors, which the BIPM does when it calculates the timescale known as Echelle Atomique Libre (EAL). The problem with EAL is that since it involves a precision of nanoseconds, it is not in phase with the standard measure of time, the SI second. As a result, EAL needs to be adjusted to coincide with the SI second in order to produce International Atomic Time (TAI). TAI is then converted into UTC by adding leap seconds.[71] Since EAL is not distributed, this adjustment does not affect the use of UTC.

Secondary Standards of Primary Importance

Those who build and manage atomic clocks recognize that the definition of the second in terms of the transition of cesium-133 uses outdated technology.[72] In a case of the definition of a standard lagging behind the technology to define it, there has been a push to develop what are called *secondary standards*.[73] These are representations of the second that use something other than cesium but are defined in relationship to the primary standard so that users can use the secondary standards with confidence. For instance, rubidium clocks achieve greater precision than cesium clocks, but since cesium is the standard for defining the second, rubidium clocks provide a secondary standard. Metrologists relate the number of rubidium transitions in a second to the number of cesium transitions that define the second. Some secondary standards have achieved levels of uncertainty that are better than the primary standard of cesium.[74] In fact, the four rubidium fountain clocks at the United States Naval Observatory have proven reliable enough to contribute to the USNO's definition of time, which then, in turn, contribute to the BIPM's definition.[75] Right now, the definition of the second is in a state where the duration of 9,192,631,770 transitions of cesium is increasingly determined by the use of secondary standards of transitions of rubidium or hydrogen.

This has led to discussions by time metrologists to redefine the second by referring to atomic transitions other than cesium, such as rubidium. This would allow a definition of greater precision than can be achieved with cesium.[76] Until that decision is made, the second remains defined by transitions of cesium-133 even as many use hydrogen masers or rubidium fountain clocks instead of cesium beam clocks. For that matter, by the time the standard changes, it is likely that the cutting edge of time metrology will have adopted a different element or even a completely different technology (say, based on light rather than on atomic oscillations) to achieve an even greater precision than previously possible. In contrast, any suggestion to redefine the duration of the second in relationship to Earth's current rotational period is met with exasperation by time metrologists. The calculated second of a perfect clock is beyond challenge; only how it is measured can be debated.

In effect, because of technological developments and the mixture of traditional ways of conceptualizing time, such as the day, conflicting with precise time measurement, which focuses on tiny units of time, there is even now a form of time pluralism. As a result, policy discussions about time standards do not just involve improvements in technology or algorithms, but also discussions and even debates over how standards should be maintained, defined, and even chosen. In their discussion of time measurement, Claude Audoin, of France's Centre national de la recherche scientifique (CNRS), and Bernard Guinot, the director of the BIH until 1985, wrote: "We must accept that we have moved into an era of multiple time scales."[77] The leap second debate discussed in the next chapter is the most public of these discussions about multiple timescales.

The Clock Users' Delusion

The effectiveness of time metrology encourages the public's lack of consciousness of time measurement. The history of the current dominant timescale is a history of an attempt to impose the semblance of order on timekeeping through algorithms and weighted averages—an order not achieved in the present, but only in hindsight. Only by noting the difference between clocks and the calculated standard the previous week can one estimate how far off the calculated standard they are now. To stave off disorder, the alternative suppresses the importance of the Earth's cycles and uses statistics. The appearance of a totalizing, standardized, global time is the result of a great deal of technoscientific effort, and not a given.

Despite time metrology, for most people the concept of the rotational day is the fundamental unit of time. They view hours, minutes, and seconds as fractions of a day. Weeks, months, and years are viewed as aggregates of days. This leads to the erroneous assumption that the units of time are fungible,[78] but they are not. Because of the leap second, some minutes are sixty-one seconds long, and some days 86,401 rather than 86,400 seconds long. In contrast, time metrology takes a definition of the second in terms of transitions of cesium-133 atoms as the fundamental unit of time.

The current day is a different kind of day from what it used to be. It is a day not tied to Earth rotation and cycles of daylight. Paul Virilio called it "The *electronic day* of telecommunications" that "is no longer the astronomical day of the ephemerides: the indirect lighting of electronic images now replaces electric lighting, just as the latter once replaced sunrise."[79] In his words, this is a "false day of indirect visibility,"[80] in that nobody sees atomic time. The day has become an example of a time unit that is independent of both social time and natural cycles.[81]

Time metrology serves the public, and this creates an inherent tension between the ideas the public has about time and the approach of time metrologists. Since much of time metrology is within government agencies or institutions established and supported by international agreements, most time metrology is both scientific and political. This fosters a tendency for metrologists to balance the scientific goal of achieving ever greater precision and the cultural conservatism of conceptualizing a day in relationship to Earth's rotation. Such a synthesis of irreconcilable cycles was bound to eventually create discussion and debate as to which timescale should be the one that is distributed globally. Those involved in the distribution of timescales argue that to broadcast two kinds of time would cause confusion.[82] Indeed, in 2023, Mario Maniewicz, Director of the ITU Radiocommunication Bureau, wrote: "There is consensus that alternate time scales and system times should not be utilized as timing reference sources."[83] The consequence of this assumption is the need to choose which timescale is the one that is broadcast. The need to choose fosters debate.

The desire to broadcast a single timescale fits with the ideas of global standards that have emerged since the Treaty of the Metre in 1875. Standards of weights and measurement are essential for science and commerce.[84] The SI second, as the fundamental unit of time, is a primary standard—a standard from which other standards are derived.

Since there are so many ways of relating atomic timekeeping to Earth rotation, and for that matter, astronomical and atomic definitions of the second, achieving a standardized timescale and units of time is a process, and cannot be taken for granted. There are simply too many factors that could lead to a proliferation of timescales, which would threaten efforts at achieving globally recognized standards. Somebody needs to make decisions and debate the wisdom of those decisions. Such is demonstrated by the debate about the leap second.

Notes

1. Banerjee and Matsakis, *Introduction to Modern Timekeeping*, 249.
2. Levine, *Geography of Time*, 82.
3. Crary, *24/7*; Birth, "Biological Consequences of Globalization."
4. Banerjee and Matsakis, *Introduction to Modern Timekeeping*, 258.
5. Arias, "Perspective," emphasis added.
6. Latour, *Reassembling the Social*, 228–29, emphasis in the original.
7. Adam, *Time*, 117; Barry, *Political Machines*, 18; Nowotny, *Time*, 93.

8. Adam, *Timewatch*, 112–14; Anderson, *Imagined Communities*, 22–24; Taylor, *Secular Age*, 208–9.
9. Peirce, *Charles Sanders Peirce, Volume 5*, 457; *Charles Sanders Peirce, Volume 7*, 341.
10. Crease, "Charles Sanders Peirce," 39.
11. Peirce, "Swinging Pendulums"; Lenzen and Multhauf, "Gravity Pendulums," 322–28.
12. Hassan, "Time and Sovereignty," 34–47.
13. Hassan, "Time and Sovereignty," 42.
14. Foucault, *Power/Knowledge*, 39.
15. Pargman and Palme, "ASCII Imperialism," 196.
16. Virilio and Lotringer, *Great Accelerator*, 83.
17. Virilio and Lotringer, *Great Accelerator*, 90.
18. Krislov, *Product Standards*; Quinn, *From Artefacts to Atoms*.
19. Galison, *Einstein's Clocks*, 85.
20. Birth, "Standards in the Shadows"; Loconto and Busch, "Standards."
21. Howse, *Greenwich Time*, 109.
22. Howse, *Greenwich Time*, 113.
23. *Daily Evening* Bulletin, "Change of Time," 1.
24. Bartky, "Adoption of Standard Time," 49–50; McCrossen, *Marking Modern Times*, 114–15.
25. Gordon, "Railway Clocks."
26. Parrish, "Litigating Time."
27. Jones, *American Ruling Cases*, 395–98.
28. Rich and Farnham, *The Lawyers Reports Annotated*, 610–19.
29. Jones, *American Ruling Cases*, 399–402.
30. Gordon, "Railway Clocks," 396.
31. Guinot, "Review of Time Scales," 19.
32. BIPM Time Section, *Annual Report*; Arias et al., "50th Anniversary"; Guinot, "Some Properties of Algorithms"; Levine, "Statistical Modeling"; Panfilo et al., "New Prediction Algorithm"; Panfilo et al. "New Weighting Procedure"; Thomas et al., "Time Scales."
33. Matsakis, Levine, and Lombardi. "Metrological and Legal Traceability," 60.
34. Whibberley et al., "Local Representations of UTC."
35. O'Connell, "Metrology," 153.
36. CGPM, "Resolution 1."
37. Arias et al., "50th Anniversary."
38. Pickering, *Mangle of Practice*, 51.
39. Ihde, *Instrumental Realism*, 80.
40. Latour and Woolgar, *Laboratory Life*; Latour, *Science in Action*; Lynch, "Representation is Overrated."
41. Boer et al., "Perspective of the Instruments."
42. Baird, *Thing Knowledge*, 68.
43. Ihde, *Postphenomenology*, 57; Rabkin, "Rediscovering the Instrument," 57; Shapin and Schaffer, *Leviathan and the Air-Pump*.
44. Whitehead, *Introduction to Mathematics*, 2.
45. Clemence, "System of Astronomical Constants," 169.
46. Clemence, "System of Astronomical Constants," 169; Smith, "International Time," 480.
47. Clemence, "System of Astronomical Constants," 169–71.
48. Scheibe and Adelsberger, "Nachweis von Schwankungen"; Danjon, "Le Temps" ; Danjon, "La Rotation"; Stoyko, "Sur l'irrégularité de la rotation de la Terre"; Stoyko,

"Sur la périodicité dans l'irrégularité de la rotation de la Terre"; McCarthy and Seidelmann, *Time*, 51.

49. McCarthy et al. "Physical Basis," 1907; Deines and Williams, "Earth's Rotational Deceleration"; Siddall et al., "Sea-level Fluctuations."
50. McCarthy, "Precision Time."
51. McCarthy and Seidelmann, *Time*, 56.
52. Clemence, "System of Astronomical Constants," 171.
53. Clemence, "System of Astronomical Constants," 169; see also Newton, *Mathematical Principles*.
54. Newcomb, *Astronomical Papers*.
55. Sadler, "Ephemeris Time"; BIH, *Bulletin Horaire*, 39–42.
56. Gingerich, *Eye of Heaven*, 20.
57. Levine, "Time and Frequency," 11.
58. Barry, *Political Machines*, 63.
59. Thévenot, "Governing Life by Standards," 797.
60. CGPM, "Resolution 1."
61. Hudson, "Commonly Used Time Scales," 820.
62. Hudson and Atkinson, "Redefinition of the Second," 31.
63. Essen, "Time Scales," 161.
64. Arias et al., "50th Anniversary," 899.
65. McCarthy et al. "Physical Basis," 1908.
66. Arias et al., "Timescales at the BIPM," S146.
67. Audoin and Guinot, *Measurement of Time*, 123–25.
68. Hafele and Keating, "Around-the-World Atomic Clocks," 169.
69. Reinhardt, "Relativistic Effects," 396.
70. Nelson, "Relativistic Time Transfer."
71. See Arias et al., "Timescales at the BIPM."
72. Arias and Petit, "Hyperfine Transition," 10.
73. Arias et al., "50th Anniversary."
74. Guéna et al., "Contributing to TAI," 108.
75. Peil et al., "Evaluation of Long-Term Performance."
76. Arias et al., "50th Anniversary"; Bize, "Unit of Time"; Gill, "Definition of the Second"; Godun, "Atomic Clocks."
77. Audoin and Guinot, *Measurement of Time*, 64.
78. Bluedorn, *Human Organization of Time*, 16–30.
79. Virilio, *Landscape of Events*, 65.
80. Virilio, *Lost Dimension*, 89.
81. Hope, "Time, Worker Exploitation, and Global Capitalism," 51.
82. Sadler, *Transactions*, 365.
83. Maniewicz, "Future of Timekeeping," 7.
84. ISO, *Economic Benefits of Standards, volume* 1; ISO, *Economic Benefits of Standards, volume 2*; ISO, *Economic Benefits of Standards*; Krislov, *Product Standards*; Quinn. *From Artefacts to Atoms*.

Chapter 7

Using Atoms versus Earth's Rotation

> Tempus est mundi instabilis motus, rerumque labentium cursus. (Time is the measure of a wobbly world, and things slipping away.)[1]
>
> —Rabanus Maurus, *De Computo*

While the creation of timescales is outside of the awareness of most clock users, occasionally there is a noticeable problem. Such problems are disruptive to the narrative of timekeeping progress and the associated pushing of time pluralism outside of public awareness. One such problem is synchronization in response to a leap second—an extra second inserted into UTC to reconcile it with Earth's rotation. On the morning of Sunday 1 July 2012, the computer systems for Qantas Airlines, Mozilla, Reddit, Gawker, and Amazon crashed. The cause was quickly identified—it was an improper implementation of the leap second.

Computer systems rely on timing to coordinate tasks and as part of their defense against malicious code. Timestamps ensure secure communications and are critical to preventing "man-in-the-middle" attacks—attacks in which a threat gets in the middle of a communication to monitor, impersonate, or disrupt the communication. These features of computer technology ideally operate outside of the notice of the computer user even as they are important concerns of systems administrators. When computer clocks meant to be synchronized significantly disagree, and one second is a significant disagreement, this fosters endless loops which can cause servers to crash.

The difference between a leap year and a leap second is that leap years come at predictable intervals that can be built into operating systems, but leap seconds are not predictable. The current standard is that the difference between UTC and UT1, a form of mean solar time, must be less than 0.9 seconds. The International Earth Rotation and Reference Systems Service (IERS) monitors the difference between UTC and UT1 and often announces a leap second when the difference reaches 0.4 seconds. The announcement gives six months' notice with the leap second being inserted on either 11:59:59 UTC on 31 December or 30 June. The six-month notice puts a burden on system administrators to develop and implement patches to prevent the leap second from creating cyber-chaos. The goal is to manage the systems so that the average user does not notice the leap second, much less experiences a problem resulting from the improper handling of this troublesome second.

Science, Technology, and Institutions

The debates over the leap second policy reveal differences in how atomic physicists and astronomers think about the definition of time standards.[2] As discussed in the last chapter, scientists disagreed on how to respond to the availability of reliable atomic clocks. That said, there were shared assumptions about the features of timescales shared by most of the people involved. Going back to what Louis Essen wrote about timescales: they had to indicate the time of day, the time of year, and provide a measure of duration.[3] There were additional assumptions based on the imagination of the ideal clock as a device displaying a timescale. First, the measure of duration had to be uniform and reliably replicated. Second, that unit must be internationally accepted. Lastly, there should only be one reference timescale defining time for the ideal clock.

In the context of these ideas, the leap second was a compromise between atomic and astronomic timekeeping. Despite this, it left the assumptions tied to the ideal clock unquestioned, as well as the idea of a single international time standard to guide all real clocks.

Another issue was implicit in these debates: who determines a global standard? With the advent of atomic clocks, there emerged a source of precise time units that competed with celestial timekeeping. The context of the competition between these approaches was the Cold War and decolonization. In this context, there emerged three distinct facets to the leap second issue: the measurement of time, the definition of a timescale, and the determination of institutional responsibility for time measurement and timescale definition.

Leap Seconds and the Measurement of Time

As discussed in the last chapter, units of time had been defined in terms of a fraction of either Earth's rotation, or when that proved to be too irregular, a fraction of Earth's tropical year. Atomic clocks offered a different possibility. Instead of defining units of time in terms of fractions of the day or year, these clocks allowed a definition of time units in terms of frequency—a count of atomic transitions. In other words, time units could be defined based on adding small increments of duration rather than by dividing long durations. This was not merely looking at the problem of defining time units in a new way, but it offered to metrology the measurement of previously unimagined levels of precision. Conceptually, one can imagine a billionth of a second defined in terms of a fraction of the tropical year, but technologically, one can measure a billionth of a second using atomic transitions.

These different approaches to time pitted astronomers against atomic physicists. It also suggested a shift in how time metrology experts imagined the ideal clock even as most clock users did not follow these debates and culturally held on to increasingly obsolete views of clock time. With the astronomical focus, Earth's cycles were the source of time measurement. With the atomic focus, the ideal clock could reliably measure fine levels of precision anywhere in the universe without reference to the Earth.

This also suggested a shift in how units of time were determined. As the introduction to the 1969 *Le Colloque International de Chronometrie* put it, clocks that achieve great precision "produce" time rather than being content to "observe" it.[4] Astronomical versus atomic timekeeping pitted those who observed and measured large durations against those who produced small durations by the counting of atomic transitions.

As argued in the introduction, clocks mediate knowledge between experts and users. A clock is a conduit for how clock users imagine time. What is at stake in the conflict between atomic and celestial timekeeping is whether the clock emphasizes the ability to achieve precision or whether the clock represents Earth's rotation. The debate between atomic physicists and astronomers is not merely over what a clock should do, but, by implication, over how to influence users' concepts of time.

With atomic timekeeping, the variation of the Earth's rotation could be very precisely measured. In 1962, the International Astronomical Union (IAU) reported, "During the past three years improvements have been made in the accuracy of time determination. A significant development was the introduction of atomic clocks for control of transmissions and for use in studying variations in the speed of rotation of the Earth."[5] In that same report, the goal for the synchronism of time signals was one millisecond.

Each advance in precision demonstrated in a new way the unreliability of the Earth day as a uniform, stable unit of time. In this conflict between privileging Earth rotation versus privileging uniformity, the choice in metrology has consistently been to favor uniformity and stability rather than representing Earth's cycles. But this does not make the Earth and its cycles go away. In fact, it lays the foundation for ongoing time pluralism because charting the Earth and its cycles remains important to many.

Efforts to Define a Global Timescale

In 1952, Commission 31 of the IAU, also known as the International Commission of Time, reported, "Astronomers are now forced to consider two different systems of time." One system, called *astronomical time,* was based on the observation of Earth's rotation. The other system, called *ephemeris time*, was based on uniform units of duration "in which the length of the day is invariable." The commission noted that "The difference between ephemeris time and astronomical time is inherently unpredictable."[6]

Resolution 7 of the 1952 meeting of the IAU adopted the ephemeris second as the unit for measuring time. This second was 1/86,400th of the length of day on 0 January 1900, as determined by Simon Newcomb.[7] In 1954, the astronomer André Danjon wrote on behalf of the CGPM, the organization that defines units of measurement, that the IAU definition of *ephemeris time* could not be considered an authoritative measurement standard without the approval of the CGPM.[8] Danjon had become the director of the Paris Observatory in 1945 after the liberation of France from the Nazis, and he was known for his work on the definition and measurement of time. Rather than defining the second in terms of Newcomb's calculations for the day, in the 1920s Danjon recast it as a fraction of the tropical year.[9] In the late 1940s, the American astronomer, Gerald Clemence also related 1/86,400th of Newcomb's calculated day to a fraction of a tropical year.[10] Based on such suggestions, Charles Volet, the director of the BIPM, urged the CGPM to come up with its own definition of the second.[11] Volet was not an astronomer, but a physicist and mathematician. He joined the BIPM soon after he finished his studies at the University of Lausanne. At Volet's urging, in 1954 the CGPM deferred the matter to the International Committee for Weights and Measures (CIPM) in the CGPM's resolution 5. The CIPM is a group of eighteen scientists who study metrology and make recommendations to the CGPM. At the same meeting, the CGPM also made the definition of the second the fundamental unit of time as part the CGPM's Resolution 6 to establish an international system of units of measure.[12]

At its 1956 meeting, the CIPM discussed how to define the second, and unanimously passed a resolution to form the Comité Consultatif pour la Définition de la Seconde (CCDS),[13] which would later change its name to the Consultative Committee for Time and Frequency (CCTF). The CGPM then charged this committee with studying time metrological technology and techniques and making recommendations on how the second's definition would be determined.

The CCDS met for the first time in 1957.[14] It was chaired by Danjon. The membership of the committee consisted mostly of astronomers with a few members, like Louis Essen, being physicists working on atomic or quartz crystal clocks. Danjon had used a crystal clock to first document the seasonal variation in the rotation of the earth.[15] The problem with crystal and atomic clocks in this time period was their long-term stability, but with the invention of improved clocks, particularly the cesium beam clock in 1957, pressure grew to define the second in terms of atomic transitions.

By 1963, such pressure led the CCDS to embark on a project to compare different types of atomic clocks to determine which was the most reliable. The committee hoped to find the best clock for determining a definition of the second based on atomic transitions. At the time, the membership of the committee was changing. The president of the committee was no longer the astronomer Danjon, but the physicist Harry Barrell, who was head of the Standards Division of the United Kingdom's National Physical Laboratory. Astronomers were no longer in the majority, either. The number of atomic physicists equaled them. The 1963 meeting focused on whether to define the second in terms of cesium or wait for additional studies of other atomic clocks, like hydrogen masers. The first vote had eleven in favor of immediately defining the second atomically using the isotope cesium-133 with twelve voting against.[16] The tide began to turn in favor of an atomic definition of the second when several physicists in attendance pointed out that time signals based on Essen's atomic clock were already broadcast by six time laboratories.[17] When the vote was taken a second time, only two voted against the motion,[18] and one of these was Donald Sadler, an astronomer who was the superintendent of the United Kingdom's Nautical Almanac Office. Sadler complained that he thought it inappropriate for the vote to be taken a second time.[19]

The Leap Second Compromise

The adoption of an atomically defined second as the International System of Units' (SI) standard only addressed the problems created by tying a unit of measure to Earth's irregular rotation. It did not change Earth's rotation, nor the astronomical definition of a day. Previous to 1972, the difference between 86,400 SI seconds and the average rotational day had been reconciled by the International Bureau of the Hour's (BIH's) introduction of "frequency changes in multiples of 50 parts in 10^{10}"[20]—a practice that has come to be remembered as the "rubber second" because it made some seconds longer than others. Many metrologists disliked the rubber second. For instance, Louis Essen, wrote: "It is illogical to apply offsets to a unit which is constant in order to make it agree with astronomical periods which are not constant,"[21] and G. M. R. Winkler, the director of the USNO Time Service, observed that "it has become evident ...that one could not continue to allow the possibility of frequency changes every year."[22]

In the pursuit of timekeeping uniformity, there was an emerging institutional pluralism. While the CCDS was discussing atomic time, Study Group 7 of the Comité Consultif International pour la Radio (CCIR) considered the question of the difference between atomic timekeeping and astronomical timekeeping. The CCIR, later renamed the ITU-R, was responsible for global broadcast standards, and its Study Group 7 focused on recommendations concerning standard frequencies and time signals. Since frequency defines broadcast signals and the number of oscillations per second defines frequency, the CCIR and its Study Group 7 had interests in the second's definition.

One proposed solution was to continue the rubber second, but this was unpopular. The second proposal was to create a coordinated continuous timescale without any steps to keep it related to Earth's rotation. A coordinated timescale represents time across multiple locations—in this case the globe. Others argued against such a timescale since it was not related to the Earth's rotational day.

Many of the members of Study Group 7 were also members of the CCDS. In July 1968 the study group held an "interim meeting" in Boulder, Colorado.[23] At that meeting, Louis Essen and Gernot Winkler suggested the leap second as a compromise between atomic and celestial timekeeping. Essen was a physicist with Great Britain's National Physical Laboratory and one of the inventors of the cesium clock. Winkler was an astronomer who was the director the Time Service as the United States Naval Observatory. The support of these two men gave the proposal an air of compromise. At its 1970 meeting, the CCDS recognized the CCIR as having a good representation of the users of a global timescale and that the CCDS definition of an atomic timescale would respond to the CCIR's work.[24] This was a compromise that gave the CGPM the power to define an internal timescale based on atomic seconds, and the CCIR the power to define the global standard coordinate timescale, UTC. In effect, this created a difference between an atomic timescale not tethered to Earth's rotation (what is now

TAI) and an atomic timescale reconciled with Earth's rotation (UTC). In addition, the CGPM defined TAI as a continuous count of seconds rather than a timescale divided into days. In contrast, UTC, with its division into days, was discontinuous because of the occasional addition of a leap second to a day. Despite the desire for a single timescale, there thus were two. To avoid conflict, the CGPM instructed that TAI was not to be made publicly available.

In 1970, the CCIR adopted the leap second solution proposed by Study Group 7[25] with an implementation date of 1 January 1972.

Many time metrologists greeted the new policy with relief. Sadler described the virtues of the new system as better addressing the demand for precision in frequency than the previous rubber second because of its maintenance of a standard interval, and noted that the new system also worked better for automatic reception of time signals, whereas the old step adjustments of 0.1 second required manual resetting.[26] In different publications, Chi and Fosque of NASA and John Henderson of the National Research Council of Canada described the new policy as an "improved Coordinated Universal Time."[27] Winkler wrote, "I think it is going to be a good system."[28] The media reports differed in their emphasis from that of the metrologists. The *New York Times* mentioned that 1972 would be "the longest year thus far in the history of recorded time."[29] The *Los Angeles Times* also fixated on the concept of the longest day on 30 June 1972: "Today is the longest day of your life," but the tone of the article was clearly tongue-in-cheek:

> But today's alteration raises some important questions.
> Will the delay create a crisis at bars beginning their happy hour?
> Will unions demand overtime pay?
> Will men develop 4:59 o'clock shadows?
> Will everyone be early for work Monday morning?[30]

Understanding the trajectory of metrological science, Humphry Montague Smith, then head of the Time Department of the Royal Observatory, offered a prophetic observation about the leap second policy: "There is bound to be continuing pressure for the adoption of AT [atomic time] without steps in much notice."[31] While Donald Sadler thought the leap second was better than the rubber second, he also viewed the atomic-based timescale associated with the leap second as the death of Greenwich Mean Time. He reputedly marked each leap second by wearing a black tie as a sign of mourning.[32] The fundamental problem with the leap second was that it attempted to reconcile two different givers of time: Earth rotation and atomic transitions. Time metrologists understood this, but the general public did not. Culture dictated the clock's display, and the display was a very limited bandwidth to communicate complex ideas. By means of the leap second compromise, the clock's display hid the time pluralism of atomic time versus terrestrial time.

Standards Organizations and the Leap Second

This debate unfolded in a period with an acceleration of the implementation of newly defined global standards. In 1960, The CGPM began to move to a new International System (SI) of standard units of measure. The goal was for measurement to no longer be tied to specific objects, such as the meter rod housed in Paris, but to the use of technology and physics to define all units of measure. Standards are concepts materialized by policies decided upon by specific institutions. As Elizabeth Cullen Dunn observes, "although standards present themselves as *episteme* ...standards need an *oikodomi*, a material context in which they are transformed into action and effect."[33] The SI system changed the material context of measurement standards from objects to replicable procedures.

The CIPM meeting in 1969 noted that there were now several organizations interested in defining the global timescale in relationship to SI units.[34] The confluence of atomic and celestial timekeeping resulted in several organizations becoming involved in the discussion over time standards: the International Astronomical Union (IAU), the International Committee for Weights and Measures (CIPM), the General Conference on Weights and Measures (CGPM), the International Bureau of the Hour (BIH), the International Bureau of Weights and Measures (BIPM), the Radiocommunication Sector of the International Telecommunication Union (ITU-R, originally the CCIR), and the Institute of Electrical and Electronics Engineers (IEEE). Paradoxically, pluralism of institutions with a stake in measurement standards is behind the effort to create a single reference timescale.

These institutions developed in parallel with slightly different priorities and with some overlap of personnel. The IAU had traditionally taken a leading role in issues of defining time by means of clocks measuring celestial cycles—a legacy of an era when astronomical observatories were the key sites for time metrological practice. Its focus was on units of time as fractions of the tropical year.

The CGPM, CIPM, and BIPM are the heirs of the Metre Convention of 1875, which created an international structure for the development and maintenance of standards of measurement. The CGPM consists of delegates from its member states and meets every four years to discuss standards of measurement. The CIPM consists of eighteen specialists in metrology elected by the CGPM. The role of the CIPM is to make recommendations to the CGPM about standards and to oversee the operations of the BIPM. The BIPM consists of a permanent staff whose day-to-day responsibility is to manage, make available, and promote the SI system of metrological standards.

Originally, time standards were not a responsibility of the BIPM but a different institution—the International Bureau of the Hour (BIH). The BIH operated out of the Paris Observatory. It was tied to the IAU and closely allied with the BIPM, and as such had participated in the trend of maintaining standards to serve increasingly global commerce and regulations.[35] Due to its tie with the IAU, the BIH consisted of astronomers. With the advent of atomic timekeeping, atomic physicists increasingly made up the membership of the CGPM and the CIPM. Once the CGPM defined the second in terms of atomic transitions in 1967, the tension between astronomical timekeeping and atomic timekeeping came to the fore. With the increasing dominance of physicists in discussions about time policy, the BIH was disbanded in 1987 with its time-keeping functions transferred to the BIPM and its astronomical functions charting Earth's rotation housed in a new institution, the International Earth Rotation and Reference Systems Service (IERS).

The CGPM's efforts were not only a matter of defining time, but part of the creation of the International System of Units (SI) for weights and measures—the CGPM created similar committees to define the meter, electrical standards, and standards for telephone communications. The goal was to define standards in a way that was no longer tied to a particular laboratory or observatory but could be replicated anywhere using standard procedures. For instance, the meter is now defined by how far light travels in a vacuum in 1/299,792,458th of a second.

The CGPM worked on defining the international system of units of measure at the same time the CCIR (later called the ITU-R) developed standards for radio, television, and telecommunications. The CGPM defined the SI second; the ITU-R/CCIR defined frequencies in relationship to the second. This led to a confluence of interests between the two bodies. Up until the 1960s, the ITU-R/CCIR had not played a role in defining timescales but instead had emerged as a body regulating the distribution of time signals. In the early days of radio, such signals played a crucial role in defining broadcast frequencies. The most famous example was the Eiffel Tower transmitting time signals tied to the BIH in Paris.

The CCIR gradually envisioned its role as far broader than regulating time signals and broadcast frequencies. By the 1960s, it increasingly saw itself as the regulatory body to determine what timescale would be made globally available. In the period when this happened, radio time signals were the primary means of synchronizing clocks across space. It was the CCIR that gave the final approval to the leap second policy. In 2000, ITU-R passed a resolution calling on UTC and leap second policy to be studied. By 2001, the CCIR had become the ITU-R and had added computer network communications to its expanding domain of broadcast signals and telecommunications standards.While the jurisdiction of the ITU-R seems to have been taken as a given,[36] and it seems to have been in place by 1969 when G. M. R. Winkler, the director of the USNO Time Service, indicated that the CCIR (the ITU-R's predecessor) would be in charge of worldwide timing.[37]

By the 1970s, networked computing started to emerge, and with it, a need for standards for computer protocols. This included the use of precise and accurate timestamps.

The International Institute of Electrical and Electronics Engineers (IEEE) sets standards for computer systems. Among these standards is POSIX, which stands for "Portable Operating System Interface," and which ensures compatibility between information technology systems. Since everything computers do involves the recording of timestamps, standardized timestamps are important for compatibility between systems. So, both the ITU-R and IEEE have an interest in setting a single global standard. Officially, only the ITU-R has jurisdiction over the global timescale, but the functioning of information technology is of major concern to the ITU-R, giving IEEE standards great influence over ITU-R regulations.

Since time synchronization was increasingly a matter of computer servers and networks, time metrologists, electrical engineers, and computer scientists developed more interests in common, fostered by the Precision Time and Time Interval meetings (PTTI), which are now sponsored by the Institute of Navigation. These are annual events that bring together metrologists, government officials, technicians, engineers, and others interested in time and frequency technologies. It made sense for PTTI to be sponsored by a navigational body because chronometry was essential to determining longitude. Even with the advent of satellite navigation systems, time was still involved, because the navigation satellites are clocks in orbit broadcasting time signals.

The tie between computer networks, time signals, observatories, and time standards created a complicated web of relationships between these different organizations. In this web, the CGPM claimed responsibility for defining the second. The CCIR/ITU-R claimed responsibility for defining the timescale used in time signals. IEEE was in charge of computer standards. The overlapping membership of their technical committees and the PTTI meetings created a network of scientists discussing and debating time policy.

The structural relationship between these different organizations is fluid, and the history of time standards bears this out. Based on atomic timekeeping, in 1967 the CGPM defined the second as 9,192,631,770 transitions of cesium-133. This was roughly equivalent to the ephemeris second as defined by the IAU,[38] but the IAU recognized the potential for timescale pluralism when it reported that the difference between ephemeris time and the atomic definition of the second might result in "two definitions of time interval in use in the future, one for astronomy and one for physics."[39] Moreover, the CGPM's taking responsibility for defining the second created friction with the IAU, particularly after the CGPM did not include a recommendation from the IAU that the definition of the second include a reference to the ephemeris second as "part of the IAU system of astronomical constants."[40] Complicating matters, the CCIR was not using UTC in 1968 when it declared itself as the governing body over the characteristics of time-signal transmissions, and consequently, the characteristics of the timescale to be broadcast.[41] As Roger Beehler of the US National Bureau of Standards noted, "the CCIR has traditionally adopted a much broader view of its scope in the time and frequency area and, in fact, has expanded its interests significantly beyond the narrow topic of standard frequency and time broadcasts."[42]

So, in fact, there were several competing institutions and organizations. They can be viewed as falling into four distinct categories with different constituencies: institutions in charge of industry standards, such as the IEEE; institutions in charge of global metrological standards, such as the CGPM and BIPM; scientific institutions, such as the BIH and IAU; and regulatory institutions, such as the CCIR/ITU-R. In a flow chart in a National Bureau of Standards report in 1974, scientific and regulatory organizations were placed under the United Nations, and the BIH, which was responsible for charting Earth rotation and maintaining time standards, was categorized as a scientific organization.[43] The IEEE was not represented in this flowchart because in 1974 the IEEE was not involved in time standards.

With the dissolving of the BIH in 1987, this weakened the relationship between those defining the second by atomic transitions, and those, such as the IAU, that had traditionally defined the second as a fraction of measured celestial phenomena. It also weakened the tie between defining the second and the CCIR/ITU-R. The weakening of these ties was great enough for Terry Quinn, who was the Director of the BIPM from 1988 to 2003, to leave the ITU-R out of his flow chart of responsibility over the setting of time standards.[44]

The change in the composition of Study Group 7 of the CCIR when it became the ITU-R in 1992 had a subtle influence on deliberations. Study Group 7 retained its focus on time and frequency and space research, including the charting of Earth rotation, but it was now an umbrella organization for four different working groups, with Working Group 7A devoted to time and frequency signals, and Working Groups 7B, 7C, and 7D devoted to various astronomical applications. The structural change weakened the influence of astronomers on time and frequency standards. The scientific constituencies represented on that committee in the 1960s were now in different working parties, and only one of these working parties was explicitly charged with time standards, even as all the working parties had a stake in these standards.

The ITU-R, and before it the CCIR, has many advisory committees, including those working on time standards. These committees are staffed by scientists, many of whom also serve in time laboratories, as well as the CGPM, but the delegates to the World Radiocommunications Conferences, which is the legislative body of the ITU-R, are not specialists in time metrology. Instead, the ITU-R is an institution within the United Nations set up to achieve international cooperation in telecommunications. As Quinn points out, its stated purpose is not the development of metrological standards.[45] Yet, even as the leap second issue was raised, the assumption was that the CCIR/ITU-R would be the body to decide the issue, not the CGPM.[46] In a bit of a sleight of hand, since the ITU-R (and its predecessor the CCIR) is responsible for the time signals that get broadcast,[47] it took on the role of being responsible for *some* time standards, namely the determination of a global timescale, but *not* for the determination of the fundamental unit of the timescale. The result was a split institutional responsibility over timescales with the current situation being that the definition of the SI second is in the hands of the CGPM and CCTF,[48] not the ITU-R, even as the leap second policy is seen to be in the hands of the ITU-R.

Currently, under the guidance of the CGPM, the BIPM is responsible for defining the SI second and calculating TAI and UTC. UTC is now the standard against which clocks are compared. ITU-R regulations recognize UTC as the globally available standard timescale. Working Party 7A of the ITU-R guides its time policies. As stated by the Joseph Achkar, the chair of the ITU-R's Working Party 7A, "As Working Party 7A's studies indicate, the determination of how the international reference time scale relates to other time sources …falls under the authority of the International Bureau of Weights and Measures (*Bureau international des poids et mesures*—BIPM) in conjunction with the Consultative Committee for Time and Frequency (CCTF), the International Committee for Weights and Measures (*Comité international des poids et mesures*—CIPM) and CGPM. However, standard-frequency and time-signal emissions and dissemination …reside within the authority of the ITU-R."[49] In 2020, a memorandum of understanding between the ITU and the BIPM outlined this division of labor, although as this document states "This MoU is not intended as, and will not be deemed or construed as, a binding agreement between the Signatories."[50] The memorandum of understanding also stated that the BIPM would have a representative on the ITU-R's Working Party 7A.

Despite this memorandum and the BIPM's participation in deliberations of Working Party 7A, there is friction because while the ITU-R's Study Group 7A consists of scientists, as an agency of the United Nations, the ITU-R is also a political body. In such a body, the desire is to seek agreement among delegates, but there are certain nations, the United States in particular, that because of their involvement in standards committees have great influence. Yet, the internationalization of standards is a case of declining authority of states over time metrology.[51] This shapes persistent and future time pluralism.

Setting Clocks—the Challenges of Time Synchronization

Before computer networks providing time synchronization and satellites sending time signals, the common way of transmitting time and frequency information involved radio transmissions—something that anybody with access to a shortwave radio could use. People would listen to these signals and adjust their clocks manually. The Eiffel Tower started to broadcast time signals on 23 May 1910, based on the clocks of the Paris Observatory. WWV, the US time signal broadcast, began in 1920. Its broadcasts consist of pings coded so that one knows the beginning of each minute and hour.[52] As long as high levels of precision were not needed, these radio signals were serviceable, but the way in which a shortwave signal travels makes its signal delays variable by up to tens of milliseconds. In 1971, Gernot Winkler predicted:

> If operators have trouble getting WWV to milliseconds—and I know many who have—then the trouble will be compounded with the new systems: they will necessitate more intensive training, there is no question about it. You may underestimate that. The application of much larger corrections will force everyone into that when in fact most people today don't bother about receiver delays, or propagation delays of 8 milliseconds or things like that.[53]

A more labor- and time-intensive method of distributing time and frequency information was literally to carry clocks from one location to another.[54] Indeed, in the early days of synchronizing atomic clocks, such clocks occupied a seat in first class during their travels.[55] But by 1971, people who relied on time standards saw this means of time transfer as awkward. For instance, Rear Admiral R. J. Schneider of the US Navy's Naval Electronics Systems Command observed that "flying clocks have become a nuisance, since airplanes are subject to hijacking and couriers carrying electronic equipment are regarded suspiciously at various borders."[56] Moreover, satellite telecommunications had reached the point where clocks riding in airplanes was no longer necessary.

In 1993, Judah Levine wrote of the then-available methods for disseminating time and frequency information that "none [of these] is likely to be completely adequate either from the point of view of adequate stability and accuracy to satisfy high-end users or from its ability to be scaled up to serve an increasing number of users with moderate requirements who are more interested in ease of use and low cost."[57] With the growth of network computing and eventually the internet, Network Time Protocol (NTP) developed by David Mills became a popular means of transferring time information that solved the problems of human error in setting clocks and negated the need to transport atomic clocks on airplanes. NTP involved a server receiving reliable time signals from several sources, dropping the outliers, and then synchronizing to an average of the remaining times.[58] Improvements to time synchronization protocols laid the foundation for the leap second problem. Before computer networks and reliance of levels of precision to a microsecond or better, precise synchronization of clocks was not a widespread problem. With advances in timekeeping precision and the coordination of computer systems across extended networks, time synchronization across those networks became a challenge that needed to be met.

Satellite navigation revolutionized time distribution by making accurate and precise timescale information available across the globe and in ways that allowed devices to steer themselves. In 2004, Thomas Parker and Demetrios Matsakis stated that GPS "delivers time with unprecedented accuracy. Free to the user, GPS is the world's most-accurate globally available one-way source of time and frequency,"[59] providing much of the globe with a version of GPS "accurate to better than 5 ns r.m.s. [root mean square]."[60] Early on, they demonstrated that GPS could be used for purposes of time synchronization to high levels of precision with, even in 1999, the time difference between the USNO's UTC and GPS time "guaranteed to be less than 100 ns," and the estimated offset "guaranteed to be accurate to 24 ns."[61] This level of accuracy and precision has made GPS extremely attractive for obtaining time and frequency information,[62] and since GPS signals are available over a large portion of the planet, obtaining such high-quality time signals only required a GPS receiver rather than building a fiber-optic infrastructure to receive internet-transmitted NTP. Increasing numbers of technologies were adopting GPS time signals as their source of time for their timestamps, such as Precision Time Protocol (PTP). It has become woven into the functional fabric of many modern technologies—so much so that it has been referred to as a "stealth utility."[63] It became clear that the availability of time via satellites would reach many more locations at many more times than time signals broadcast across telephone lines, fiber-optic networks, or broadcast by means of shortwave radio. The more systems that relied on precise timekeeping, the more systems that had to be adjusted every time a leap second was announced. The more connected systems were to each other, the more likely an improper leap second adjustment in one system would have a negative cascading effect.

The leap second compromise did not eliminate the astronomical or atomic timescales, but, in a sense, institutionalized their time pluralism. Both timescales were needed to know when a leap second had to be implemented. This meant that the coordination of atomic clocks produced an atomic timescale, TAI, and astronomers observed a different timescale based on mean solar time, UT1. With the leap second policy, UTC relied on both TAI and UT1.

As long as UTC was a matter of policies governing broadcasted time signals, the number of institutions broadcasting it was relatively small. With the advent of networked computing and information technology, the number of devices and systems synchronizing according to programmed protocols increased into the hundreds of millions. The clock as a cognitive tool had expanded from devices that kept time, but required users to set them, to devices that kept time that was automatically synchronized to other devices. Now, rather than users being aware of how well or poorly their clocks perform, users have come to assume that their clocks are reliably synchronized simulacra of an ideal clock. Each leap second was a threat to synchronization, and revealed that the ideal, free-running, uniform, precise, and accurate clock did not exist, and that UTC was a compromise in response to the time pluralism of atomic and celestial timescales. Rather than choosing a timescale, the leap second compromise left time pluralism intact, but with a hope that the average clock user would not notice. In fact, as the importance of precision has increased, and a one-second step seems increasingly large, implementation of leap seconds has spawned additional time pluralism. Some institutions (e.g., Google, Facebook, Alibaba, and Microsoft) do not follow the protocol for inserting a leap second but instead engage in what has come to be called a *leap smear*. In a leap smear, small fractions of a second are added over the course of several hours rather than an entire second being added after 11:59:59 UTC. When a leap second is added, leap smears create their own time pluralism since there is no standardized protocol for them, leading Patrizia Tavella, the director of the Time Department of the BIPM to acknowledge, "An inserted leap second is labeled as 23:45:60—a clock-time unforeseen in most modern, digital systems. This discrepancy caused the proliferation of ad hoc methods that are increasingly being used as alternatives to the leap second."[64]

Notes

1. Rabanus Maurus, *De Computo*, column 676.
2. See Nelson et al., "The Leap Second"; Finkleman, "Future of the Leap Second,"; Seago et al., *Decoupling Civil Timekeeping*; Seago et al., *Requirements for UTC.*
3. Essen, "Time Scales," 161.
4. *Le Colloque International de Chronometrie*, "Introduction," 5.
5. Sadler, *Transactions*, 361.
6. IAU, "1952 General Assembly," 461.
7. IAU, "1952 General Assembly," 66; see also Sadler, "Ephemeris Time."
8. CGPM, *Comptes Rendus des Séances*, 91.
9. CGPM, *Comptes Rendus des Séances*, 93.
10. Clemence, "System of Astronomical Constants"; "Revising the Official System."
11. CGPM, *Comptes Rendus des Séances*, 62.
12. CGPM, *Comptes Rendus des Séances*, 80.
13. CIPM, Procés-verbaux des séances, 78.
14. CCDS, *Procès-Verbaux de Séances.*
15. Danjon, "Le Temps" ; "La Rotation."
16. Essen, *Revolutions in Time*, 306.
17. CCDS, "3rd Session," 62.
18. Essen, *Revolutions in Time*, 306.
19. Essen, *Revolutions in Time*, 306.
20. Winkler, "UTC Worldwide Adjustment," 387.
21. Essen, "Time Scales," 164.
22. Winkler, "UTC Worldwide Adjustment," 387.
23. CCIR, *XII Plenary Assembly, Volume VII*, 182.
24. CCDS, "5th Session," s20.
25. CCIR, *XII Plenary Assembly, Volume III*, 258.
26. Sadler, "Coordinated Universal Time," 33–34.
27. Chi and Fosque, "A Step in Time," 82; Henderson, "Foundation of Time and Frequency," 487.
28. Winkler, "UTC Worldwide Adjustment," 387.
29. *New York Times*, "Two Seconds," 29.
30. *Los Angeles Times*, "Wait a Second," D1.
31. Smith, "International Time," 485.
32. Sadler, "A Personal History," note by editor, 160.
33. Dunn, "Standards without Infrastructure," 118.
34. CIPM, *58th Session*, 16.
35. Quinn, *From Artefacts to Atoms.*
36. Beard, "Role of the ITU-R."
37. Winkler, "Summary of Comments," 1.
38. IAU, "Commission de l'Heure," 362–63; "Resolutions Adopted by the General Assembly," 41.
39. Sadler, *Transactions*, 365.
40. IAU, "Time," 344.
41. ITU, *Radio Regulations*, 408.
42. Beehler, "Role of the Consultative Committee," 323.
43. Barnes and Winkler, *Standards of Time and Frequency*, 14.
44. Quinn, "The BIPM," 895.

45. Quinn, "Time," S124.
46. See Chadsey and McCarthy, "Relating Time," 243.
47. Beard, "Role of the ITU-R"; Fosque, "The International Radio Consultative Committee."
48. Arias et al., "50th Anniversary," 901.
49. Achkar, "Progress on Time Signals," 13.
50. BIPM/ITU "Memorandum of Understanding," 5.
51. Johns, "Temporal Rivalries."
52. See G. Nelson et al., *NIST Time*; Walls, "The Standard Frequency."
53. PTTI, "Panel Discussion," 397.
54. Allan et al., "Precision and Accuracy."
55. NBS, *Atomic Clocks.*
56. Schneider, "Time and Frequency," 12.
57. Levine, "Frequency Dessemination," 577–78.
58. Mills, *Computer Network Time Synchronization.*
59. Parker and Matsakis, "Time and Frequency Dissemination," 32.
60. Parker and Matsakis, "Time and Frequency Dissemination," 33.
61. Levine, "Introduction," 2,583.
62. Levine, "Time and Frequency Distribution," and "Time Transfer"; Tullis, "Screwed in a Millisecond."
63. Parkinson and Powers, "Fighting to Survive," 18.
64. Tavella, "Coordinated Universal Time," 17.

Chapter 8

Time Pluralism and the Leap Second Problem

...for technology is a campaign against the day.[1]
—Dietrich Bonhoeffer, *Creation and Fall*

Hundreds of millions of devices rely on time synchronization for their accuracy, but the cutting edge of time metrology is precision. E. Felicitas Arias, who was Director of the Time Department of the International Bureau of Weights and Measures until her retirement in 2017, observes that: "The main problem with the leap second is that it is not compatible with time synchronization with a precision better than one second."[2] In effect, the leap second is an awkward compromise between the accurate representation of Earth rotation and the desire to measure ever smaller units of time. It is awkward in another sense—when it not done correctly it reveals that time is the product of policy decisions that sometimes have unexpected consequences. Time metrologists seek a smooth and unproblematic functioning of timescales, not publicly noticed disruptions. The leap second elicits among them a discourse of "failure," "problems," and even a fear that people might die as a result of a botched leap second implementation.

In 2013, Dr. Dennis McCarthy, by then retired from overseeing the United States Time Service, warned that "failure to provide a source of continuous time and frequency information particularly could result in potentially dangerous or expensive problems in the future."[3] This was not long after the debacle of the 2012 leap second in which multiple computer systems crashed, including Qantas Airlines, Reddit, Mozilla, Gawker, and some Amazon platforms. These crashes provided evidence of the threat the leap second posed to synchronization. If the leap second policy did not end, McCarthy saw a future in which the leap second created problems larger than what happened in 2012.

Some, mostly astronomers, disagreed with McCarthy. They felt strongly that the globally disseminated timescale should be coupled to Earth's rotation. This then led to a debate about the risks of keeping leap seconds versus the risks of ending the policy, particularly with regard to how inserting a leap second sometimes disrupted air traffic control. An exasperated McCarthy remarked, "I would not want to travel in the air when the leap second is being inserted."[4]

Few know the leap second policy or its consequences as well as McCarthy. He was in charge of the time service at the United States Naval Observatory (USNO). He is the author of numerous articles on time metrology and time standards. With P. Kenneth Seidelmann, he wrote the definitive history of the emergence of our current time policies and practices.[5] He was among the first people to recognize the dangers of the leap second for wireless communications in his article with William Klepczynski titled "GPS and Leap Seconds: Time to Change?"[6] All of these were reasons to listen carefully to his arguments. But what grabbed my attention—something not captured in the transcripts of his remarks—was the frustration in his voice when describing the current situation. Despite his warnings, which he had in fact been giving since the late 1990s, the leap second policy was still in place in 2013 (and it is still in place in 2025, although now set to be phased out by 2035), and as he stated, "It is so easy at this point just to do nothing until a disaster occurs."[7]

On the other hand, the possibility of a disaster, as opposed to a mere occasional inconvenience, might be overstated.[8] There have been twenty-seven leap seconds inserted between 1 January 1972 and 2024. They have caused disruptions but no life-threatening disasters. Because of this, some argue that the risk posed by leap seconds is exaggerated, and that the costs of decoupling UTC from Earth's rotation are greater than the dangers of keeping the leap second in place.

There is a new aspect of the problem, however. When the CCIR/ITU-R approved the leap second policy and implemented it in 1972, Earth's rotation appeared to be slowing. Every leap second since the first involved adding a second. Networked computer systems and current operating systems were designed assuming that leap seconds would only ever be added, not subtracted. But now Earth's rotation is speeding up, leading to the possibility of a negative leap second—basically creating a single fifty-nine-second minute to reconcile UTC with Earth's rotation. Writing code to remove a second is a different and possibly riskier matter from the well-established practices of leap second computer patches. Moreover, there are those who think that since the design of many current systems only made it easy to add leap seconds, a negative leap second would create significant problems.

Particularly because of the risk of a negative leap second, a disaster could occur. When the IERS announced a leap second in 2015, CNN stated that it threatened to "break the Internet,"[9] and *USA Today* predicted "computer chaos."[10] We live in a computer-reliant world, and the functioning of computers depends on timing and accurate timestamps. Incorrect insertion of a leap second or neglect to insert one leads to desynchronization and causes system failures. Power grids, telecommunications, financial markets, and transportation are just some of the systems dependent on reliable time synchronization. By the late 1990s, with the advent of the internet and increasing reliance on synchronized computer networks, the irregular insertion of a leap second posed a problem,[11] and with the growing possibility of a negative leap second, fears of widespread computer outages have increased.

The current demand for precision combined with increased network connectivity brought the conflict between Earth rotation and atomic timekeeping to the fore, placing time metrologists in the position of doing science, communicating science, and suddenly engaging in political processes about time policies with those who have little understanding of metrology. It also revealed a split between those who wanted the standard timescale to be tied to Earth's cycles and those who were seeking an entirely abstract, uniform timescale that would hold true anywhere in the universe.

Such are the stakes, but most clock users are not aware of what the scientists are debating. Moreover, because of the complicated institutional structure discussed in the last chapter, it is not entirely clear what organization is responsible for the leap second. Traditionally, it was thought to be the ITU-R, but in November 2022 the CGPM passed a resolution calling for the tolerance between Earth rotation and atomic timekeeping to be increased by 2035, thereby making leap seconds less common. Shortly after, a small chorus of those interested in the debate pointed out that it was possible for the ITU-R to not go along with the CGPM resolution, although at its World Radiocommunication Congress in late 2023, the CGPM proposal was passed without controversy.[12] Neither institution answers to the other, and both rely on overlapping membership on their committees for their policies to be coordinated.

Technology has created what Manuel Castells calls a "network society"[13] and cultivated what David Harvey calls "time-space compression."[14] The leap second is a product of the policies and structures that enable global information technology. To understand the current debate and how decision-making has stagnated in the ITU-R, it is useful to put the leap second into the context of global information technology connections.

Consequences for the Networked Globe

It is remarkable to think that a debate that involves a handful of scientists ends up determining time standards for the globe, but because of the way in which information technology is networked, this is the case. That is an immense responsibility that those scientists take seriously. Those who rely on the global timescale UTC span many different cultures and nations to include all users of synchronized timepieces. Contrary to Frederic Jameson's idea that postmodernism displaces time,[15] UTC is the material dissemination of a uniform time standard that shapes many aspects of social life across space and cultural boundaries.

The transcultural desires for stability, accuracy and precision combine with specific cultural and national attitudes. Since nobody knows all these cultural variations, debates about time policies involve a great deal of effort among the small number of time metrologists in imagining what different constituencies desire. That said, reliable information is only available for a few technological constituencies like astronomers, computer systems administrators, precision clockmakers, and time metrologists. The LISTSERVs on which some of these debates unfolded had ongoing discussions of what Jews, Hindus, and Muslims would think of UTC being decoupled from Earth, but such discussions were mostly speculative. The cultural diversity of clock users and their clocks is thus hidden from time metrology just as time metrology is hidden from the majority of clock users. Clocks mediate this relationship with minimal interaction between producers of timescales and the users of the timescales.

At the moment, the measurement of precision is the cutting edge of the cultural demands placed on time metrology and metrologists' efforts—so much so that the USNO Time Service was renamed as Precise Time. This idea did not come from the time metrologists but reflected the interests of those who did not grasp everything that the Time Service did. It is a feature of the digitalization of how time is imagined and materially manifest in the shift from analog clock displays to digital ones. This digitalization involves the choice of representing fractions of a second using a decimal system and is essential to the timestamps that form a critical component of network protocols. In fact, the shift from using a sexagesimal system for seconds and minutes to a decimal system for fractions of a second is a mostly unexplored element of the history of timekeeping. Conceptually, this was made possible by the adoption of decimal time by France in the 1790s. Then in the late nineteenth century, Frederic Winslow Taylor used decimal fractions of a minute in order to assist in his time and motion studies.[16] In scientific research using small fractions of a second, devices such as the Hipp chronoscope or Bergström pendulum chronoscope were adopted,[17] and several sciences started to focus on the increment of one tenth of a second.[18] These devices used decimal fractions of a second and were precise to one thousandth of a second. By the early twentieth century there was a movement to adopt decimal time for scientific purposes, but only some divided the second into tenths. The decimalization of fractions of a second only became the standard when, in 1960, the second became part of the International System of Units (SI) in which all measures were divided decimally.[19] Digital displays of time lent themselves to decimalization, and together, they allowed the representation of ever-finer levels of precision.

Social, economic, and political processes that rely on networked computing embrace the move from analog to digital representations. This increases tension between the pace of human-guided social processes versus the pace of digital processes.[20] While social scientists focus on the issue of speed,[21] telecommunications reached the limits of speed decades ago[22]—signals cannot travel faster than the speed of light. When information technology reached the limits of speed, the value of precision increased. In fact, while analog approaches to time can support speed and acceleration, digital representations are essential to precision. Digital displays can easily represent tiny fractions of a second—it is just a matter of adding more decimal places after the zero. On the other hand, an analog clock indicating nanoseconds would need to have one billion discernable increments for each second, and consequently, 60 billion for the entire dial. Even if these increments were spaced at .25 millimeters, they would add up to a dial with a circumference of 15,000 kilometers and a diameter of around 4800 kilometers. To achieve ever greater levels of precision, the ideal clock must be digital.

Precision creates the leap second problem as much as Earth's foibles do. When the greatest level of precision clocks could reliably achieve was at the level of several minutes, the difference between clock time and rotational periods would take centuries to notice. Now, with levels of precision that can be as fine as a quintillionth of a second (i.e., eighteen decimal places), the variable rotation of Earth is abundantly apparent and precisely charted. The greater the precision achieved, the greater the irregularity that is noted in the cycles of which humans are aware. The greater the precision, also the more clunky an irregular change of one second appears.

Time-Space Compression and Timescale Pluralism

The social scientific approach to telecommunications and globalization takes as a technological given what is the result of complex processes that mix politics, science, technology, and even nationalism. This mix of technological progress and cultural desires is something that frames debate and forces choices to be made. Often expediency drives the choices around the world, which is why GPS has become so important—it is readily accessible almost everywhere on the planet.

Global connectivity raises the stakes of global time standards. David Harvey's idea of time-space compression[23] and Manuel Castells's network society[24] are only glimpses of the consequences of the infrastructure that allows for such almost-instantaneous global connections. Time-space compression rests upon a global time infrastructure that involves debates, discussions, policy-making, and deliberative decision-making about the time standards that define the time displayed by clocks. Its consequence is the creation of forms of chronopolitics in which those with the power to impose concepts of time use them for purposes of maintaining power and accumulating wealth,[25] and lurking in the shadows playing large roles are satellite navigation systems, particularly GPS.

In a purely physical way, time-space compression fosters the very time pluralism that time metrologists see as a challenge. There are several causes of this. First, no clock indicates the correct UTC time with certainty. Technology breaks, and those who manage it are acutely aware of this fact. For this reason, the effort at creating redundancies results in no clock serving as the perfect clock, only close simulacra of the imagined perfect clock. Second, since UTC is calculated, one only knows how well a clock performs *after the fact*; in other words, after UTC has been calculated and distributed. Third, in synchronization of clocks, one must account for signal delays. No signal can go faster than the speed of light, and signals that travel by cable are even slower. Fourth, as discussed earlier, since the Earth rotates, by the time a signal travels from one point to another point, that point has been rotated to a new location. If the signal comes via a satellite clock, then that satellite's clock can run faster or slower as its orbit takes it further and closer to the Earth due to gravity's influence on general relativity. For all these reasons, relativity must be taken into account. Satellite navigation systems do not work well unless this is done correctly.

In the world of the distribution of time signals, the simple notion of time-space compression belies the complexity of the mathematics behind it and obscures the significance of tiny increments of time. When financial-trading algorithms are developed based on data feeds with nanosecond precision,[26] or when electronic surveillance technology utilizes nanosecond timestamps to get a very precise sequence of communications traffic, the relationship between two locations on the globe does not allow instantaneous communication, but a complicated physics problem requiring sophisticated algorithms for things to function properly. This is not a matter of time-space compression, but of time-space coordination.

The globe is the problem, not in the sense of it being erased in thought as a result of telecommunications as some, such as *New York Times* columnist Thomas Friedman, have suggested,[27] but in the sense that it is ever present and disrupting calculations. As more and more technology moves about the solar system rather than fixed to the Earth's surface, the problem grows worse. The problem with UTC and the leap second was that both were the result of coordinate systems of time within an Earthbound frame of reference and a definition of the prime meridian based on the location of an observatory built by a colonial power. With increased precision and dreams of the perfect clock time, there increased the need for a timescale that was not bound to Earth. Time metrology had to confront the effect of relativity on the creation of a reference timescale. Relativity not only requires considering time dilation in calculations but also forces decisions about frames of reference. One of Einstein's insights was that once one grasped the theory of special relativity, which assumed an inertial frame of reference, then one could grasp the reality that every frame of reference in the universe is moving at various rates in various directions in relationship to all other frames of reference.[28] From this point of view, there is a problem in adopting an Earthbound frame of reference, particularly for imagining a future of space travel. In effect, there is a need for timescales that are not tied to Earth.[29] So astronomers and metrologists recognized multiple timescales, as well as the problem of having timescales too focused on our planet. But even acting on this knowledge would not solve the problems inherent in UTC.

The Problem with the Leap Second Compromise

Spurred by such problems, in the late 1990s articles started to appear by notable time metrologists making the argument that the leap second policy needed to be evaluated and possibly eliminated. McCarthy and Klepczynski pointed out that whenever a leap second is inserted, there is a period in which synchronization disappears: "while a leap second is being introduced, and until synchronization is established, communications can be disrupted between some systems."[30]

In April 1999, the Consultative Committee for Time and Frequency (CCTF, formerly the CCDS) met, and one of the items on the agenda was a report (labeled CCTF 99-18) by Dennis McCarthy on the leap second. McCarthy documented "several causes for possible concern," which included the "communications problems" and "annoyance of people in charge of systems disseminating time; and consequent proliferation of independent time scales, not including leap seconds, for specific purposes" (e.g.,

GPS time).[31] He then provided the following list of possible responses to these problems:

- maintain the status quo;
  discontinue leap seconds in UTC;
  Pro: would be supported by those in charge of disseminating time.
  Con: unlimited growth of [*UTC* — *UT1*]
- redefine the second;
  Pro: fundamental solution;
  Con: would require definition of other physical units; the solution is only temporary and its efficiency is not certain.
- Increase tolerance for [*UTC* — *UT1*]
  Pro: easy to accomplish;
  Con: date of adjustment unpredictable, difficult to establish acceptable limit.
- Periodic adjustments of UTC at larger intervals:
  Pro: date of adjustment predictable;
  Con: number of leap seconds unpredictable, larger discontinuities[32]

In the discussion that followed his presentation, there was agreement that the CCTF did not have the authority to eliminate leap seconds, and that Terry Quinn, then-Director of the BIPM, should write a letter "in collaboration with Dr. McCarthy" to the bodies that might have such jurisdiction: the International Astronomical Union (IAU), the International Earth Rotation and Reference Systems Service (IERS), the Radiocommunication Sector of the International Telecommunication Union (ITU-R), and unspecified "navigation bodies."[33]

Soon after the appearance of this article, Chadsey and McCarthy presented at the 32nd Annual Precise Time and Time Interval Meeting called "Relating Time to the Earth's Variable Rotation."[34] This presentation pointed out that "as technology advances, the time steps required to maintain that level of synchronization [between Earth's rotation and atomic time] become more inconvenient for some users to implement."[35] The discussion that followed favored eliminating the leap second, but Ronald Beard of the US Naval Research Laboratory pointed out that the ITU-R would feel responsible for defining a new policy, and that the process of getting the ITU-R to approve any change would be protracted and complicated.[36] Beard would become president of the ITU-R's Working Party 7A, which was charged with studying the leap second issue and making a recommendation to the ITU-R. Despite the challenges he foresaw, he argued that the change is necessary in order to serve the growing number of applications that need a continuous time reference and that are disrupted by the leap second.[37]

In 2009, at another meeting of the CCTF, there was another discussion of the redefinition of UTC to eliminate leap seconds.[38] McCarthy provided a summary of the history of UTC. Most time laboratories favored ending the leap second policy. For producers of a uniform timescale, the goal was to implement the ideal clock as the metrological standard for timekeeping. In their minds, such a clock would not be subject to the annoying foibles of terrestrial cycles.

There was resistance to these proposals. Some scientists argued to keep the leap second in order to keep the global standard timescale tied to Earth rotation.[39] There are reasons for this, particularly for those who chart the movement of objects in the sky—whether those be astronomical objects,[40] human-made objects like satellites,[41] or asteroids that threaten Earth.[42] The problem astronomers face is that the Earth rotates, so observational technology, such as telescopes, needs to move with the rotation of Earth in order to stay focused on the desired object. This is most easily accomplished by a clock drive that moves the telescope to compensate for Earth's rotation. Satellite navigation also requires a close relationship between Earth's rotation and timekeeping. With GNSS systems critical to global time synchronization, this requirement of satellite navigation could not be ignored. Decoupling the standard reference timescale, UTC, from Earth rotation would undo an automatic relationship between clock drives and Earth rotation and consequently require frequent adjustments to those drives.

Opposition also came from the United Kingdom and Russia. The United Kingdom realized that decoupling UTC from Earth rotation would result in UTC no longer being tethered to the Greenwich meridian. Russia was against the change because its satellite navigation system, GLONASS, represents Earth's rotation (UT1) as an offset from UTC, and its broadcast values cannot handle the offset values greater than a second. If the leap second was eliminated, the difference between UT1 and UTC could grow to many seconds, or, eventually, minutes. This would destroy the viability of GLONASS.

Who would define time for the ideal clock? As long as time was a matter of material clocks, nations could control their devices. Once it became a matter of the clock as an abstract standard that was produced and controlled, the question of who would have power over that standard came to the fore. This is probably why Judah Levine pointed out in the discussion that followed McCarthy's paper "that the issue of leap seconds is mainly political, not technical."[43]

Based on his experience, Dennis McCarthy felt strongly that trying to maintain the charade of reconciling mean Earth rotation and atomic timescales was pointless, and in the long run, counterproductive. The situation had been made worse by timescale pluralism coupled by an increasing number of Global Navigation Satellite Systems (GNSSs) with their own timescales. About this development Włodzimierz Lewandowski, who was the Principal Physicist in the BIPM's Time Department, said, "Confusion reigns."[44] GNSS systems consist of high-quality clocks in space that keep their own timescales. These timescales differ from UTC in several ways. First, the European (Galileo), Chinese (BeiDou), and United States (GPS) operate GNSS systems with timescales that do not include leap seconds. Only the Russian system, GLONASS, uses leap seconds. Second, these GNSS timescales are steered to "tick" in time with UTC, so that they can be easily converted into UTC by GNSS receivers. Third, with the exception of Russia's GLONASS, which matches UTC, GNSS timescales involve continuous counts of seconds from a specified date. In the case of GPS and Galileo, the beginning of the timescale is 0 hour 00 seconds UTC on 6 January 1980; in the case of BeiDou it is 0 hour 00 seconds UTC on 1 January 2006.[45] Satellite navigation systems were material clocks that some viewed as a threat to the goal of achieving an unequivocally agreed upon and universal clock time standard.

Of these systems, GPS has the greatest global usage.[46] In 1980, the USNO successfully tested it as a means of time transfer by the USNO.[47] Since 1996, when it was made available for civilian purposes, it has been used for non-military time synchronization[48] Like most other GNSS systems, GPS time is not UTC. The USNO steers it to represent UTC.[49] While GPS clocks do not implement leap seconds, GPS signals indicate the difference between GPS time and UTC as kept by the USNO. In effect, GPS is de facto its own timescale that serves as a reference timescale that gets converted to UTC. Because of its global reach, many users of precise timekeeping around the world prefer to use GPS, making this American military technology an essential part of global technology.[50] As a result, for most of the world the time of the ideal clock has come to be a function of the United States' military. This is not just a matter of clock performance, but the distribution and availability of information for clock synchronization.

Postcolonial Reticence

The leap second debate is unbalanced. This is part of a larger aspect of time-space compression that is insufficiently explored. Through control over time standards and telecommunications technology, a small number of scientists representing a small group of nations influences the entire world through the definition of the clock time standard. It involves identified individuals making arguments in journals, conferences, and on a LISTSERV dedicated to the topic, but the majority of the users of material clocks, particularly those outside the English-speaking world, are without a published voice in the debate. To time metrologists and policy makers, these users are an aggregate of cultural interests who can only be imagined, and who are not elevated to the position of participants in the discussions. At best, they are given a collective voice that unduly homogenizes them and results in the technocratic participants in the leap second debate speaking for populations based largely on speculation.[51]

Robert Hassan emphasizes the unevenness of digitality, which time-space compression enables.[52] One of the crucial sources of difference Hassan identifies is between the extraction of capital through material processes versus through digital relays of information. The latter are more individualized than the capital and supply chains of the former.[53] So while capital accumulation seeks to subsume the globe under a single uniform virtual space and time, global uniformity is not the case—the colonized have a different view than the colonizer,[54] and the aftermath of decolonization has not led to global equality but to clear stratification in the ease or difficulty in which different groups have access to telecommunications and are able to travel without restrictions.[55] Moreover, with regard to regimes of time, there is a shift away from the Eurocentric modernity of colonialism to a global modernity.[56] Time-space compression, then, and the time-keeping standards that go with it, are not homogenizing influences on modernity, but instead they create differences,[57] and consequently time pluralism.

This creates its own problems. At the ITU-R's World Radiocommunication Conference (WRC), the careful technical and scientific arguments that are meant to shape policy become submerged in a gathering of delegates more focused on issues other than time metrology. The WRC meetings have an agenda, most of which has to do with issues like broadband accessibility. No matter how hard experts in time metrology tried to make their cases, even holding conferences dedicated to the leap second issue in preparation for WRCs, the structural reality has been a gathering of delegates from over 180 nations, many of whom are not involved or interested in time metrology but who at least put up a façade of lacking enough information to make a decision. With the decline of colonialism, international bodies such as the ITU-R become sites of policy inertia rather than global authority. The inertia that emerges out of the multiple readings and misreadings of a policy recommendation becomes a form of postcolonial resistance to Western scientific institutions, such as the BIPM and its parent CGPM, proposing the policy.

Amidst the diversity of views is hostility toward former colonizers and global cities. Colonialism has ended, but the domination of a few cities over global life, including the creation of UTC, remains. Economic power is concentrated in a small number of metropoles, such as New York, London,

Tokyo, and Hong Kong, and these are also central nodes in telecommunications.[58] A small number of time laboratories of inordinate importance supply data for the creation of UTC. This is due to the algorithms for the weighting of atomic clocks in the calculation of UTC. The weights for contributing clocks are determined by their long-term stability, but metrologists recognized that if the calculation relied on the most stable clocks, then only a small number of clocks in a few wealthy nations would dominate the calculation. As a result, the maximum weight any one clock can be given is four divided by the number of clocks used in the calculation.[59] The BIPM tested the robustness of this algorithm in the calculation of rapid UTC (UTCr)—a version of UTC that is released weekly—more frequently than *Circular T*, which is released monthly. The test involved removing 20 percent of the contributing clocks from the calculation, with the normally heavily weighted USNO clocks being specifically mentioned as removed.[60]

In addition, computer technology has cultural biases—since the design of computer components and code emerged in the Euro-American context, they contain deeply embedded Western biases,[61] such as timestamps using Gregorian calendar dates. Coupled with these biases certain parts of the world have an inordinate amount of influence over the UTC calculation. A graphic example of this is found in Lewandowski et al.[62] In this article, figure 1 shows the international time links reporting to the BIPM in 2004. In this figure are fifty-six time laboratories. Of the fifty-six, six are in North America, twenty-six are in Europe, sixteen are in Asia, and only eight are in South America, Africa, and Oceania combined. At the time of Lewandowski et al.'s article, the National Metrology Institute of Germany (PTB) and the BIPM were the central nodes receiving reports from every time laboratory. The BIPM weighted some laboratories very heavily in the determination of UTC based on the number and quality of their atomic clocks. These included the United States Naval Observatory in Washington, DC (USNO), Japan's National Institute of Information and Communications Technology in Tokyo (NICT), Time and Frequency Laboratory of Besançon in France (LTFB), China's National Time Service Center in Xi'an (NTSC), the Polish Time and Frequency Laboratory (PL), and the United States National Institute of Standards and Technology in Boulder, Colorado (NIST). Most nations have either no time laboratory or a small one that contributes marginally to the production of UTC.

The uneven distribution of time metrological expertise and time laboratories combined with nationalist ideologies and postcolonial resentments makes the ITU-R and its World Radiocommunication Conference sites of passive disagreement. These are contexts where those nations with stakes in creating UTC and defining the standard clock time argue, but most delegates and votes at any WRC belong to those nations without time laboratories. Such passive disagreement gives the illusion that the driving force behind changes in time standards are the nations with the greatest technological contribution to the determination of UTC. This reflects what Dipesh Chakrabarty identifies as the common assumption that the "driving engine" in late capitalism resides in Europe and North America,[63] but the obstacle that passive disagreement creates in a global legislative body seems to be able to stop the "driving engine" in its tracks.

At WRC-2012, the resolution to eliminate leap seconds was on the agenda. To try to predict how delegates would vote, members of Working Party 7A, the ITU-R committee that was in charge of evaluating the leap second policy, spoke with delegates before the vote, and what they learned was that the support for eliminating the policy was not strong enough to warrant bringing the resolution up for a vote. To the consternation of those on both sides of the debate, a large number of delegates were noncommittal. The World Radiocommunication Congress of 2012 passed resolution 653, which put off the decision to the 2015 World Radiocommunication Congress, and which called for the ITU-R "1) to conduct the necessary studies on the feasibility of achieving a continuous reference time-scale for dissemination by radiocommunication systems" and "2) to study issues related to possible implementation of a continuous reference time-scale."[64]

By 2015, little further study had been done, but more debate had taken place. What followed was a great deal of polemic in the guise of studies, and studies that made little pretense of having decided their conclusions before conducting their research. Such was the case with the report to the United Kingdom's National Measurement Office in 2014. The then-minister of Science and Universities had gone on record as opposing the abolishing of the leap second before the report was even contracted.[65] The research was then contracted to a private research firm, which quickly realized that most of the public had no idea what the leap second was. As a result, the research protocol had to include educating research participants on the leap second, and in the process, shaping their views on the topic. It is no surprise, then, that the conclusion of the study echoed the sentiments of the minister who commissioned it.[66]

The period between 2012 and 2015 involved very little empirical research but a great deal of posturing. The 2015 meeting of the World Radiocommunications Congress determined that still more study was needed to generate a report for the WRC in 2023.[67] Or as the online publication *Computerworld* put it: "Tick. Tick. Tick. Clang! That was the sound of an intergovernmental conference kicking the leap-second can down the road."[68]

This stalling to make a decision led to frustration at the BIPM. The then-director of the BIPM's Time Service, E. Felicitas Arias, described the

2015 WRC meeting as "a missed opportunity to dismiss the leap second once and for all."[69] Such frustration led to those in favor of eliminating the leap second policy to stop turning to the WRC for a decision, but for the BIPM to take matters into its own hands and assert its authority over UTC by eliminating the leap second. Meetings of the CGPM determine BIPM policies, and in the November 2022 CGPM meeting it passed a resolution that "the maximum value for the difference (UT1-UTC) will be increased in, or before, 2035."[70] This does not eliminate the leap second but would make leap seconds less frequent than they presently are. Under the original leap second policy, the maximum value of difference between UT1 and UTC was 0.7 seconds, but this new resolution was meant to allow the difference to be much larger than that. The resolution also called on the CIPM to work with the ITU and other organizations to determine what the maximum difference should be. Yet, there are clear reservations about this solution. Judah Levine of the US National Institute of Standards and Technology and Patrizia Tavella and Martin Milton of the BIPM argued that this solution would have the benefit of making leaps less frequent than the current policy, but it also had two undesirable consequences. First, it increased the risk of systems administrators not having experience implementing these less-frequent leaps. Second, the larger change would have even greater impacts than the present one-second difference. Even noting these problems, these authors end up suggesting that "enlarging the tolerance is a wise provisional solution."[71]

The outcome of the 2023 WRC was different. The resolution was not for the ITU-R to eliminate leap seconds, but instead for the ITU-R to "cooperate" with the efforts of the BIPM, CIPM, and CGPM "to define a new maximum value for the difference between UT1 and UTC and on the implementation date for continuous UTC, possibly in 2035."[72] This resolution passed easily.

Timescales and Power

The original CCIR recommendation was for there to be a single global timescale, but the inherent dilemma is that the Earth's rotation and atomic timekeeping are two fundamentally different sources of time information for defining timescales. The leap second obscured the difference, but the difference remains. The leap second problem is the result of seeking to create one timescale out of two.

At many points in this now over two-decade-long debate there has been a related argument about the problems created by having multiple coexisting timescales. The claim has often been made that multiple timescales create confusion, so if it was decided to maintain both a leap-secondless timescale and a timescale coupled with Earth's rotation, it would cause problems. The underlying logic is that the clock time standard should have no competitors. As early as 1999, those advocating the elimination were writing, "Failure to provide well thought-out plans is likely to lead to a chaotic increase in the number of nonstandard time scales, resulting in confusion and a disservice to users."[73] In June 2016, at the "Science of Time" conference at Harvard University, I heard exchanges and debates from those associated with GPS, the US National Institute of Standards and Technology, the International Bureau of Weights and Measures, and other time metrology experts over this very issue. In this exchange, someone observed that the precision-timing equipment made by Meinberg, a major manufacturer of precision clocks, has a switch to allow users to change between GPS time and UTC, but the timestamps generated by computers using Meinberg equipment make no distinction between these two timescales. For most users this makes no difference, but for high-precision users, there are differences between the two timescales. GPS keeps its own system time with no leap seconds. GPS broadcasts contain as a digital correction an estimate of the exact difference between GPS time and UTC(USNO). The result is a simulacrum of UTC(USNO),[74] which itself is a simulacrum of UTC. In effect, the time on a GPS receiver is a simulacrum of a simulacrum.

As a strategy document produced by the Consultative Committee for Time and Frequency (CCTF) points out, "There is currently a proliferation of GNSS times with different relative offsets, most differing from UTC by changing and unpredictable values."[75] This concern about the proliferation of GNSS times is not a problem for most clock users because they tend to use a single GNSS system, but it could be a problem for those who combine systems. Current technology creates time pluralism even as policies seek to eliminate it.

The effort to adopt a uniform, universal timescale out of a structural muddle of loose coordination between regulatory, scientific, and standard-setting bodies has fostered the debate about the leap second. It has also been fertile ground for competing institutions defining different timescales; at the same time, Western scientists and regulators agree that there should be one reference timescale. Even with this agreement among regulatory, scientific, and standards institutions, there is no global agreement that there should be a single reference timescale, particularly outside North America and Western Europe. At a conference on the leap second in 2015, China's lead scientist on time metrology, Dr. Han Chunhao, argued in favor of eliminating the leap second while also maintaining a global system to distribute Earth orientation data, in effect, creating two different timescales. In some ways, there is already timescale pluralism behind the use of UTC.

It is not possible to know how close a local realization to UTC is in the moment—only after the fact once *Circular T* is published. As a result, every time laboratory can have its own representation of UTC.

It might seem that the timescale pluralism that is the context of the leap second debate is a creation of this mixture of science and politics. Issues of time pluralism have probably been around as long as people have attempted to reckon time—material clocks have never cooperated with an ideal of a continuous and uniform timescale. Classical antiquity provides many examples of attempting to relate solar cycles and lunar cycles to each other;[76] and time-keeping tools that display multiple timescales are more common than modern tendencies toward uniformity would incline most people to believe. The efforts to define and impose a single timescale are efforts at exercising power, but maybe a different approach needs to be taken, namely, understanding timescale pluralism. Timescale pluralism is seen as a threat to the image of a clock that charts an absolute, authoritative and uniform timescale, but its consequences are poorly understood, and material clocks are of great importance to their users and owners even when these clocks foster pluralism. As the cases in previous chapters demonstrate, timescale pluralism has been common throughout history.

Notes

1. Bonhoeffer, *Creation and Fall*, 27.
2. Arias, "One More Second," 1,178.
3. McCarthy, "Considering Time-Scale Requirements," 312.
4. McCarthy, "Considering Time-Scale Requirements," 318.
5. McCarthy and Seidelmann, *Time.*
6. McCarthy and Klepczynski. "GPS and Leap Seconds."
7. McCarthy, "Considering Time-Scale Requirements," 315.
8. Allen, "Planes will Crash!"
9. Goldman, "This Year's Y2K."
10. Griffin, "Computer Chaos."
11. See Wallace, "Software for Timescale Applications."
12. WRC-2023, "Resolution 655," 398–400.
13. Castells, *Rise of the Network Society.*
14. Harvey, *Condition of Postmodernity*; "Between Space and Time"; "From Space to Place."
15. Jameson, *Postmodernism*, 156.
16. Kanigel, *One Best Way*; Taylor, *Principles of Scientific Management.*
17. Bergström, "A Type of Pendulum Chronoscope"; Edgell and Symes, "Wheatstone-Hipp Chronoscope."
18. Canales, *Tenth of a Second.*

19. See Kovalevsky and Quinn, "International System of Units," 801.
20. Hassan, "Analogue Time."
21. See Rosa, *Social Acceleration*; Scheuerman, *Liberal Democracy*; Sharma, *In the Meantime*; Virilio, *Lost Dimension*; Virilio, *A Landscape of Events*; Virilio, *Speed and Politics*; Virilio, *The Great Accelerator*; Virilio and Lotringer, *Crepescular Dawn*; Wajcman, "Life in the Fast Lane?"
22. Angel, "When Finance Meets Physics."
23. Harvey, *Condition of Postmodernity*; "Between Space and Time"; "From Space to Place."
24. Castells, *The Network Society*.
25. Huebener et al., "Intersection of Time and Globalization," 246; Huebener et al., "Introduction," 8; Porter and Stockdale, "Strategic Manipulation."
26. Birth, "While the West Sleeps."
27. Friedman, *The World is Flat*.
28. Einstein, *Relativity*, 52–53.
29. Guinot and Seidelmann, "Time Scales."
30. McCarthy and Klepczynski, "GPS and Leap Seconds," 50.
31. BIPM, *Consultative Committee*, 102.
32. BIPM, *Consultative Committee*, 102–3.
33. BIPM, *Consultative Committee*, 103.
34. Chadsey and McCarthy, "Relating Time."
35. Chadsey and McCarthy, "Relating Time," 237.
36. Chadsey and McCarthy, "Relating Time," 245.
37. Beard, "Role of the ITU-R."
38. McCarthy, "Coordinated Universal Time."
39. Allen, "Telescope Systems"; Finkleman et al., "Future of the Leap Second"; Malys, "Proposal"; Seaman, "Systems Engineering"; Kaplan, "Time Scales"; Seidelmann and Seago, "Time Scales."
40. Allen, "Telescope Systems."
41. Seago and Storz, "UTC Redefinition."
42. Seaman et al., "Keeping Time."
43. McCarthy, "Coordinated Universal Time," 21.
44. Lewandowski, "Global Navigation Satellite Systems."
45. Lewandowski and Arias, "GNSS Times."
46. Johnson et al., "GPS SVN59 Satellite Anomaly"; Parkinson and Powers, "Origins of GPS," 30.
47. Putkovich, "USNO GPS Program."
48. NAVSTAR, *GPS User Equipment Introduction*, 11.9.
49. NAVSTAR, *GPS User Equipment Introduction*, 147.
50. Birth, "Time Lords"; Rooney, *About Time*.
51. see Spivak, "Can the Subaltern Speak?"
52. Hassan, *Uncontained*, 99–100.
53. Hassan *Uncontained*, 99–122.
54. Massey, *Space, Place and Gender*, 147.
55. Sassen, "Spatialities and Temporalities," 222.
56. Hope, "Conflicting Temporalities," 64.
57. Massey, *Space, Place and Gender*, 148.
58. Sassen, "The Global City."
59. Panfilo, "Coordinated Universal Time," 31.
60. Panfilo et al., "New Weighting Procedure," 291.

61. Friedman and Nissenbaum, "Bias in Computer Systems"; Pergman, "Reflections on Cultural Bias."
62. Lewandowski et al., "Evaluation of Uncertainties," 282.
63. Chakrabarty, *Provincializing Europe*, 7.
64. WRC-12, "Resolution 653," 1.
65. Gayle, "Britain Battles U. S."
66. OPM Group, *Leap Seconds Dialogue.*
67. ITU, "Coordinated Universal Time (UTC)."
68. Sayer, "8 More Years."
69. Arias, "One More Second," 1,178.
70. CGPM, "Resolution 4."
71. Levine et al., "Towards a Consensus," 8.
72. WRC-2023, "Resolution 655," 400.
73. McCarthy and Klepczynski, "GPS and Leap Seconds," 53.
74. Lewandowski and Arias, "GNSS Times," S220; Parker and Matsakis, "Time and Frequency Dissemination," 33.
75. CCTF, "CCTF Strategy Document," 25.
76. See Stern, *Calendars in Antiquity.*

Chapter 9

Unsettling the Clock

> The real hope for man lies not in promises of splendor or in virtuosity but in the revelation of original and authentic rhythms within the gloomy paradox of a world.[1]
>
> —Wilson Harris, *Tradition, the Writer and Society*

In February 2023, the European Space Agency stated that the Moon might need its own reference timescale.[2] Media headlines then transformed this to the Moon needing its own time zone.[3] Time zones are defined in relationship to UTC, but the Moon poses a problem. As the European Space Agency pointed out in its original report, because of the effect of gravity and velocity on time, clocks on the Moon run fifty-six microseconds faster per day than Earth's clocks. With the nanosecond precision of UTC, such a difference would make it difficult to link lunar time to UTC. This is why the European Space Agency argues for a distinctive reference time for the Moon. Culturally, however, understanding that the Moon needs its own reference time is difficult because it highlights time pluralism—hence the distortion of the original statement found in some media reports. The uniformity of the imagined ideal clock is powerful.

Synchronized clocks promote ideas of uniformity in timekeeping. While that is a goal of time metrology, even the cutting edge of atomic timekeeping recognizes that clocks differ in performance. This is why the BIPM produces the current globally distributed timescale, UTC, by a calculation rather than referencing a single, high-quality clock. Individual clocks cannot be relied upon to achieve the uniformity associated with the image of the ideal clock. Still, the information technology infrastructure of modernity depends on the assumption of the close relationship between the UTC timescale, the imagined ideal clock, and material clocks. This assumption is not false, but it relies on the robustness of clock synchronization, not on the quality of individual clocks. Synchronization is the means of all the clocks, regardless of their quality, to represent a time that is a reasonable image of the perfect clock.

Synchronization creates a disturbing language. Those who manage clocks work on techniques to "discipline" their clocks. This discipline is achieved by "slave" software to identify "errors" and bring the "slave" clocks into line with "master" clocks—clocks that are high-quality sources of time. These inherited terms make some in the precision-timing community cringe: for instance, when Demetrios Matsakis was head of the Time Service at the United States Naval Observatory, he prohibited the use of the terms, and IEEE has changed this terminology in its standards to TimeTransmitter and TimeReceiver,[4] and recommended against the the use of "slave clock" and "master clock" terminology in all technical terminology and communications.[5] Many people prefer the term "steering" to "disciplining." Still, the old terms remain commonly used. Some couple this infelicitous language with evaluations of some of the "slave clocks," particularly those in consumer electronics, as "poor performing," or "unreliable."

Due to such unreliability, frequent steering is a necessary part of synchronization. Without it, things go awry. This was demonstrated during the First Gulf War. A Patriot antimissile battery in Dhahran failed to hit its targets because of a clock error. To predict the location of incoming Scud missiles when they came into range, Patriot batteries use time and velocity calculations. The battery in Dhahran had operated for one-hundred continuous hours without synchronizing its clock with GPS, making it hopeless in protecting the base. As a result, an Iraqi missile got through the defenses and killed twenty-eight American soldiers.[6]

The longer the battery operated without the computer system being reset, the greater the error grew—to the extent of 20 percent in eight hours of continuous running.

The steering of devices to achieve uniformity through technological links is a general symptom of modernity. This replaces the cosmopolitanism of early modern Europe. Uniformity is an article of faith for most clock users, but such blind faith leads to trouble when an "undisciplined" clock is given as much credence as a faithfully synchronized clock. For Fernando Wood, temperance activists, and those who sold liquor, the slippage could amount to minutes. Today, it might be nanoseconds, but in the world of big data analysis or high-frequency trading, slippage on the order of nanoseconds can still leave the true sequence of events unclear.

In the conflicts they embody, material clocks are sedimentations of the past. Max Weber observed: "An inanimate machine is mind objectified."[7] If this is the case, then clocks are the objectification of a variety of minds spanning the millennia over which clock-based timekeeping emerged. Clocks as cognitive tools reproduce the knowledge and ideas of the past—they involve the dead thinking for the living. The alchemy of Egyptian, Babylonian, and French Revolutionary ideas combines with technological innovations discussed in chapter 6. Our timekeeping is the result of several centuries of technical innovation. Consequently, clocks are memory devices that code the knowledge of past thinkers into a device so that present users can tap that past knowledge without knowing it. In *Matter and Memory*, Bergson makes the point that memory combines with perception so that present experience makes sense. Bergson's argument about perception emphasizes the experience of the world of objects, but the idea of memory focuses on the individual,[8] not on technology. He does not explore the possibility of memory being in the world as opposed to the mind, even as he produced an object (a book) that would do just that. If we view clocks as repositories of memories about reckoning time, then the ability to make sense of time using the clock is not a fusion of personal memory with present experience, but a materialized cultural memory with present experience. This sedimentation of past desires for material clocks, such as reckoning time in relationship to Earth's rotation, clashes with current desires for uniformity. It creates a dissonant web of meaning in which the imagined, ideal clock stands for counts of atomic seconds for metrologists, but for Earth rotation for many clock users. The different interpretations of the same metrological representation mean that while clocks mediate the relationship between experts and users, steering clocks has not completely steered clock users.

This dissonance leads to the question: what if that cultural memory is not uniform, but a diversity of times embedded in the object? The cases in the chapters show that the clock is a medium between a myth of homogeneity and an unruly set of conflicting ideas. After all, from Rita Felski's feminist perspective "time is a concept of enormous complexity, including questions of measurement, rhythm, synchronization, sequence, tempo and intensity. It spans the personal and the public, work and leisure, the instantaneous and the eternal, intimate relations and global structures, everyday life and conditions of extremity. It exists at many different levels and is experienced in radically divergent ways."[9] So why should the clock privilege uniformity and standardization?

The Martiniquan philosopher Édouard Glissant wrote: "Sameness is sublimated difference: Diversity is accepted difference."[10] This general principle can be applied to the efforts to create, maintain, and even impose a single timescale on the entire globe. Modernity is often associated with a single timescale as defined by an unproblematized and fictitious perfect clock, but other timescales exist, although they are often portrayed as marginal.[11] By the twentieth century, the concept of a uniform timescale became "one of the West's most successful exports."[12] With the auto-synchronizing capabilities of digital technology, the export has become even more deeply embedded in daily life than previously when people had to at least seek a time source to which they set their watches. When it comes to clocks, we no longer choose our "masters"—they are given to us.

Clocks and Power

Time represented by clocks is not just a social construction—it constructs sociality. This construction relies on material clocks to represent a perfect clock that does not exist. As such, it is part of a growing number of standards that guide human activity in a way that leads to the "construction of certain societies,"[13] but as is the case with the relationship of a meter stick with the standard of the meter, the material objects cannot be counted upon to be absolutely faithful. Even if the clock in one's device is not very reliable, frequent synchronization with time sources creates the illusion that it is, but it is not the clock that is reliable in this case, but the synchronization protocol. Such protocols give clocks authority. Whose authority? Who controls the "master" clocks?

The power over clocks is a subtle power over the minds of clock users. The current time standard, UTC, is a bureaucratic tool in regulation; it is a legal tool in enforcement, a self-disciplining tool in daily planners, and a surveillance tool in tracking devices in electronics. Computers timestamp actions and documents, communications are timestamped, and people arrange their lives according to the clock. These subtle forms of power are crucial for the commodification of labor.[14] UTC is so deeply and widely enmeshed in our lives that, for many, it is difficult to imagine time as anything other than what devices indicate.

Such dependence on a device leads Bernard Stiegler to argue that as a species we are succumbing to systemic stupidity,[15] or as James Scott more diplomatically puts it, "high-modernist designs for life and production tend to diminish the skills, agility, initiative, and morale of their intended beneficiaries."[16] But what skills do clocks diminish? To what stupidity are we succumbing? Are we moving towards ignorance and idiocy? The power of the ideal clock as mediated by clocks depends on suppressing knowledge of the diversity of interests that have guided the designs and uses of material clocks.

Suppressing Diversity

The ubiquity and power mediated by clocks renders other temporalities invisible.[17] Invisibility is not a state of nonexistence, but one of being outside the field of vision. The image of the ideal clock makes the timekeeping of the West ubiquitous, thereby making the other temporalities harder to recognize without attention to the time-keeping tools and techniques that embody these temporalities. It also casts them as different from the ideal. While asking someone on the street about time elicits the time of day, interrogating the object the person consults then opens up a Pandora's box of diversity and debates.

There are those who suggest that the West's obsession with time measurement is exaggerated,[18] but I think this hides the power and significance of time measurement in relationship to governmentality, and this forecloses alternative ways of thinking about time that have been part of the design of clocks. It also obscures the diversity of ways in which clocks capture thought, such as in the interesting contrast of the clock being used as a model for phenomena in psychophysics and chronobiology—the metaphors of brain and body "clocks"[19]—versus the social sciences and humanities representing clocks as "unnatural." Sometimes, scholars attribute features to the West that are general features of bureaucratic governmentality. It is not only the West that has been concerned with the measurement of time—most cultures that have developed centralized bureaucracies have interests in time measurement. Notable examples include imperial China,[20] the classical Maya,[21] and the Ottomans.[22] Building monumental timekeeping devices is a common feature of empires.[23] In China, "harmony between time and space was the foundation of rulership and a unified country."[24] For the Maya, "political organization was rooted in cosmology, religion, and temporal cycling, themselves inextricably bound."[25] Within the Ottoman Empire, before the use of mechanical clocks, there were professional astronomers known as *muvakitt* who determined time and made it available from buildings called *muvakitthanes*. These timekeeping professionals adopted the use of clocks after they were introduced.[26] Seemingly, standardized timekeeping is not a natural outcome of measurement, but a result of the consolidation and dissemination of power.

What makes the Western approach different from the Chinese, Maya, or Ottoman approaches to time is not the obsession with measurement, but with uniformity as a guiding principle. A logic of standards of measurement is necessary for global capitalism. Adam Smith alluded to this in his *Wealth of Nations*: "Equal quantities of labour, at all times and places, must have the same value for the labourer."[27] Standards transcend cultural differences. This lends itself to Marilyn Strathern's idea that "a world made to

Euro-American specifications will already be connected up in determined ways."[28] Yet, contrary to this sort of narrative of the West obliterating other points of view, the standardized, uniform timekeeping that is accessory to global capitalism has not absolutely triumphed over local practices—alternatives still exist,[29] and temporal pluralism doggedly remains.

The blind faith in clocks depends on the standardization of the representation—it is this that builds faith in the standardization of measurement. By this, I do not mean forcing a choice between analog and digital displays, but that whatever the display of time, it is based on UTC. As pointed out in the introduction, most people receive little education about time and cannot imagine clocks other than those taught in grade school. In contrast, as the examples of medieval York, Renaissance Nuremberg, and Habsburg cosmopolitanism showed, clocks can vary in their form and the timescales they display. Moreover, clocks can be used in a counter-hegemonic fashion, such as in 2014, when the clock on the Legislative Palace in Bolivia was redesigned to run counterclockwise. The point the Bolivian government made was that the idea of "clockwise" adopted the movement of a shadow on a sundial in the northern hemisphere, and Bolivia is a country in the southern hemisphere. The counterclockwise clock was thus a symbol of indigeneity and resistance.[30] This redesign was unveiled at the Group of 77 meetings (an organization consisting of 134 developing nations) at which the delegates were all given counterclockwise clocks,[31] although since these were still clocks, critics pointed out that they were still derivative from European colonialism.[32] The clock is a representational medium. It can and has been used to promote time pluralism or it can be used to impose time uniformity.

The current, standardized, uniform timescale emerges out of a mediating cycle of measurement, reporting, calculation, averaging, comparison, and weighting of inputs from hundreds of atomic clocks. These steps obliterate diversity of location, rhythm, and culture. According to Jonathan Crary, by attempting to override the rhythms of life as it unfolds in particular locations, UTC creates a machine-like 24/7 global pseudosociety[33]—"pseudo" because standards become a substitute for sociability by means of making certain privileged communications easy and computer mediated even as they narrow the cultural bandwidth of communication to only what the standards permit. Stiegler sees the autonomous and calculated aspect of 24/7 society as a form of hyper-control that creates "'individualized' normativity."[34] Neither Crary nor Stiegler inquire into whose 24/7 time it is. UTC is, by definition, universal (at least until somebody settles on the Moon!), but uses of material clocks are specific events in particular places.

Science and Clocks

Science is not our salvation from this situation. The reliance of scientific research on clocks gives that research legitimacy even as it grants clocks authority. Some scientific instruments are epistemologically important,[35] and clocks are among them. Clocks are not just about how we know increments of time or time of day; they shape how we know other things. Any protocol or procedure that involves knowing duration or points in time is shaped by the use of clocks. Yet, instead of such use of an instrument leading to a concern over the epistemology of time, the instrument is critical to establishing epistemological legitimacy. In studying the diversity of phenomena, the uniformity of the timescale scientists use constrains their imagination. In science, accurate and precise timekeeping is methodologically necessary. To "know" when something takes place in science, one must have faith in the global, uniform, synchronized timescale. This legitimacy led to nineteenth- and twentieth-century concerns over individual differences in the time of observations—a problem referred to as the "personal equation."[36] That experience inherent in human experience has been replaced by a reliance on instrumentation. In many ways, the work of time metrologists is not just to pursue precise and accurate timekeeping, but to cultivate and nurture the faith of those who believe in UTC. Clocks enable certain insights by means of their uniformity and standardization, even as they cultivate ignorance by occluding other temporal possibilities.

The sophistication of time measurement today makes it an essential tool in biological, psychological, and environmental research. In the history of science, new discoveries are often a result of new instruments,[37] so privileging a single way of representing time possibly occludes discoveries. Contrary to the view of "one time fits all" that Bartky documented in his book on the movement for global time uniformity,[38] for purposes of scientific research, timescales need to be appropriate for the research conducted. Giambattista Vico's perspective on science helps to frame this problem. He posed a distinction between *verum* and *certum*. *Verum* is knowledge through knowing causes (*per causas*).[39] As Isaiah Berlin describes this, "According to this principle we can be said fully to know a thing if, and only if, we know why it is as it is, or how it came to be, or was made to be, what it is, and not merely that it is what it is, and has the attributes it has."[40] Such knowledge is only possible for that which one makes, not for that which one observes. Consequently, *verum* is true in a formal sense—humans create the rules for establishing what is true and false and what is true adheres to those rules. The units and timescales of time metrology are examples of *verum* in that their definition is established. In this sense, units of time and timescales are products of naming.[41] They are what John Austin called "illocutionary acts" also known as "speech acts"—things done with words.[42] Time metrology manufactures rather than observes a timescale, and others design synchronization protocols.

The measurements of phenomena to indicate those units are examples of *certum*, i.e., certainty. *Certum* is the epistemological status of what one chooses to observe rather than produce. In current precise time measurement, the choice is to count atomic transitions and transform the counts by means of mathematics. Because of the mathematics for calculating UTC, uncertainty creeps in not through the algorithms for creating UTC but in the comparison of UTC as calculated with the representation of UTC at individual time laboratories.[43] UTC is *verum*; the output of individual clocks is *certum*.

For many who produce knowledge, this *verum* that is UTC is the default timescale. They know no other alternative, and UTC then shapes the choice of what to observe. Contrary to Heidegger, who sees technology as a means of revealing nature,[44] in this case, technology can foreclose revelation by placing its users into a cognitive rut. Take, for example, the study of animal activity cycles. Representing activity cycles in relationship to what entrains them provides useful information. Representing activity cycles in relationship to the timescale UTC is less useful since most activity cycles entrain to something other than a clock. For instance, one study of baboons represents them as being active on average from 0900 hours to 1700 hours—in other words, baboons work from nine to five.[45] This obscures variations in baboon behavior shaped by weather and seasonal variations in daylight. Aside from how the average obscures behavioral variations, this clock time, as a version of UTC, is a distorted view, when the activity cycles should really be represented in terms of a phase relationship to sunrise and temperature—the environmental variables that seem to entrain baboon activity and which are more consistent with what is known about the relationship of activity patterns and temporal ecological niches.[46] Clocks thus elide the diversity of temporalities in the world even as clocks enable comparisons useful in creating knowledge.

There are benefits, then, to uniform clocks, but there are also those who worry about the costs. Leroi-Gourhan feared a dystopian future in which the "domestication" of time turned humans into entirely socially conditioned organisms, like ants or bees.[47] Stiegler suggested this possibility with the idea of the object that embodies knowledge (of which clocks are one example) becoming a crutch that destroys knowledge.[48] To what extent can clock users think about time without clocks, or, possibly about creating alternative clocks?

Grade School Knowledge

Maybe Regiomontanus and Walther were at an advantage in pursuing their research. Living as they did in a place in which the awareness of multiple timescales was part of daily life, they made informed decisions about the timescale best suited for their research and design instruments to indicate that timescale. Today, when for many people, the last class they ever have which is devoted to the topic of time is part of a unit on how to read a clock in primary school, there is less sophistication and awareness of choices even among scientists than there was in Regiomontanus's time.

The cultural blinders are very powerful. They limit how to imagine time. Clock time is an arbitrary signifier and the time on any clock is a disciplined signifier referring to an arbitrary signifier, but that semiotic claim does not work well with one's employer if one works for an hourly wage. Yet, with enough knowledge of timescale pluralism, and under the right circumstances, one can use that knowledge to one's advantage. Fernando Wood did this at the 1859 New York Democratic Party convention, and this was a tactic in the enforcement of liquor laws in the nineteenth century. These examples show that knowing the alternatives to the hegemonic timescale can be a source of power, because the limited distribution of the knowledge of alternatives to it gives some actors privileged knowledge. One could argue that, through technologies of synchronization, the current UTC timescale is robust enough to prevent this, but that is not the case. It is a representation that includes levels of precision too small for a human to perceive. The difference between nanoseconds is a matter of faith in technology, not of observation. How does one know that a timestamp is slightly incorrect? Since timestamps are a matter of faith, as long as one has faith in the system giving the timestamp, one views the timestamp with certainty even as the metrologists recognize its relationship to UTC as having some uncertainty.

Maybe it is not blind faith but naive faith—a faith cultivated in childhood that is never relinquished. The baggage of an early primary school level of understanding about time reckoning is a limiting factor in many scholarly disciplines outside of physics, astronomy, and time metrology. Education deeply inculcates the image of an ideal and perfect clock into children's minds,[49] and it becomes difficult for adults to shake this idea. The childlike faith in the clock gained in those early years of schooling privileges what Benjamin calls "homogeneous, empty time."[50] There is probably no intellectual pursuit emerging out of the European intellectual tradition that is immune to the grip this childlike faith in a timescale has over how scientists conceive research and carry it out. We conflate the measurement of time with time itself—an anxiety that has dogged the study of time from at least as early as Aristotle's *Physics*[51] through the present.[52]

This is suggestive of what Deleuze called a "Society of Control."[53] He contrasts such societies with those based on discipline, such as what Foucault described in *Discipline and Punish*.[54] Societies based on discipline involve external power engaged in training and organization of their subjects who conform because of a combination of the training and the sense of being surveilled. In societies of control, dependency on technology results in technology exerting subtle control over its users. The control consists of limiting awareness of possibilities, in this case, awareness that time can be represented in many different ways. Part of the control is the imposition of standardization to occlude individuality and diversity.

From Conscious Regulation to Invisible Algorithms

In the introduction, I pointed out that the question "What time is it?" typically produces the astonishing answer of somebody looking at their clock and indicating the current time. It is astonishing in many ways. Regardless of the background, gender, age, educational level, and class of the person asked, if they have a functioning smart phone, they will respond in the same way offering the same clock time. If one gathers a group of people, they will give the same answer. When considering the diversity of human behavior across cultures and periods, the current uniformity is shocking. Yet, we are not shocked. What should shock has become mundane. This new mundane experience of timekeeping shapes senses of work, space, and self that are technologically enabled and culturally and historically unique.[55]

If one asked the same question in medieval York, the responses would have been quite different. The respondent might have referred to the time of a particular institution, such as "according to the clock at All Hallows Pavement," or "according to the bells of the Abbey." If one referred to a Habsburg gift clock, the responses would have been quite different: one might have been asked if one preferred the time in temporal/canonical, Italian, great, Greek, Astronomical, or small hours. Conversion between hour systems was commonplace for Renaissance and early modern European intellectuals who corresponded with peers throughout Europe. The astronomer might have responded differently to the question than the priest, and the laborer with a still different answer. A person's timescale could be related to the person's gender, occupation, class position, and location—in short, to the person's identity and the conscious awareness of how a timescale shaped that identity.

In that situation, those interested in time knew the source of their information. The late- medieval regulations in York always identified the source of time. There was also recognition that clocks varied in performance, even as Charles V in his retirement fruitlessly tried to synchronize his clock collection. Time was tied to location. Temporal hours varied with longitude and latitude. Clock hours varied with longitude. Gift clocks intended for natal astrological charts indicated the specific latitudes.

Not so anymore. There has been a shift from knowing one's source of time and that source being local to not knowing the source of time and time being thought to be uniform globally. Wherever one travels, the timescale is the same and the "local" time is that of a time zone tied to UTC by a publicly known offset. Regardless of culture, gender, or wealth, we learn to think of clocks as uniform.

Equally astonishing is the extent to which the naturalized assumptions about time are misinformed. Most clock users think that the movement of the Sun determines the time—a quaintly Ptolemaic idea if one thinks about it, since the Sun's apparent motion is the result of Earth's rotation. Instead of conscious awareness of who provides time, those who provide the time are hidden from view, and their algorithms and practices are different from what clock users think. The fusion of clocks with cognition about time, then, creates an oddly misinformed set of people across cultural and social differences who share a set of uniform ideas without reflection.

Maintaining Misinformation Technology

In this strange, transcultural context of global misinformed uniformity, time metrology plays a role not only in ensuring the quality of the timescale, but in maintaining the misinformed uniformity. This is made clear by the leap second debate. This debate is fundamentally about the desire for time metrology to remain invisible. Public recognition of the global system of time metrology only emerges when there is a problem, such as when computer systems crash due to an improperly implemented leap second patch. The two sides of the leap second debate argue the same point. Those who want to keep the leap second argue that if the policy ends, then people will notice a decoupling of Earth's rotation from UTC. Those who want to end the leap second policy argue that if the policy continues, there will be more and more inconvenient computer system failures, some of which might be extremely dangerous. While the parties who argue do not admit it, they share the same goal—for time metrology not to attract notice.

They couple the goal of invisibility with tongue-in-cheek jokes about being "time lords." They recognize the importance of a robust, stable timescale for the smooth functioning of technology worldwide, and consequently, are aware of the importance of their work. This creates a technocratic ethos of being invisibly important and importantly invisible. A Habsburg gift clock typically had the name of its maker inscribed on it. The UTC displayed on one's computer or smart phone is anonymous.

Rather than viewing time metrology as a product of the West, it is better to view it as a sedimentation of imperialisms. After all, UTC is a cobbled-together pastiche of ancient Egyptian, Babylonian, Roman, Christian, and French Revolutionary ideas mixed with an astrologically defined week that probably comes from South Asia and is now dependent on binary computer code with its concept of zero derived from Indian mathematics to function. The hyperprecision uses Arabic numerals to display the French Revolutionary decimal system. While the current form emerged with European colonialism, in the big picture, it is the product of millennia of ideas moving from one culture to another and shaping new ideas. This is not mere diffusion, but what Fernando Ortiz would call "transculturation"[56]—a transformation of cultural knowledge resulting from the interaction of ideas from different cultural groups. To view UTC and the current timescale as European in origin erases a multicultural global set of contributions, but every material clock displays these contributions for everyone to see. It is better to view the United States, Europe, and China as having disproportionate roles in managing UTC.

So, the lack of choice in timescales is the result of a slow, transcultural sedimentation over centuries. It was not steered by a clear logic toward a known outcome, but instead, each decision leads to new challenges and debates. What emerged is a system that relies on most users lacking any understanding of timescales—as people dependent on their disciplined devices. It is a system that elides differences. The system excludes many questions that do not fit it. The privileging of the globally uniform suppresses persistently plural local exceptions.

For all this hegemonic power, it is also a fragile system. As time metrology pursues greater levels of precision, and as information technology demands greater precision in network coordination, the need to make choices has accelerated. It took centuries for the Roman Catholic Church to revise its calendar. It took decades from the emergence of railroad time to the formalization of a prime meridian and time zones. It took a few years from the invention of the atomic clock to atomic time becoming the standard of measurement. It took a few years to realize the need for a leap second, and even as that decision was made, there were those who said it was the wrong decision. The pace of having to make choices about timekeeping seems to be accelerating as precision increases. This acceleration is even as the global population of time users remains largely unaware changes in time standards. Time metrologists' efforts to improve the performance of their technology complicates their efforts to prevent problems or even disasters. All the while, they have ceded decision-making to World Radiocommunication Conferences in which many delegates reflect the general public's limited knowledge of timekeeping, who do not heed time metrologists well, and who view the West, particularly the United States, with suspicion. What presently keeps the unsettling nature of time pluralism at bay is this alchemy of postcolonialism, sublimation of difference into uniformity, technoscience, and ignorance.

The problem is with the imagination of a perfect clock, not material clocks. Clocks have been, and can be, as distinctive as the diversity of human rhythms and relationships. While the ideal clock is currently the dominant hegemony, it still relies on actual, material clocks. For this reason, it should be viewed as what Deleuze and Guattari describe as a *plateau*—a region of intensities and multiplicities without a specific goal.[57] This might seem strange in light of the histories of clocks that follow a set of implicit unilinear evolutionary assumptions, or the policy goals of time metrologists. Yet, the historical evidence reveals the evolutionary assumptions as false—there have been many clocks and there still are many timescales. Moreover, the leap second debate reveals that the interests of metrologists are not homogeneous and are easily frustrated by other forces. The ideal clock represents a strange eddy in the history of human timekeeping—a product of the confluence of globalization, colonialism, capitalism, and science. It cannot be expected that all these forces will continue to flow in the same direction.

A Ruling Timescale?

Can a single timescale adequately represent all the rhythms and cycles in the cosmos? There has not been such a timescale, and there is not presently. It is unlikely that there ever will be. Currently, while UTC is the dominant timescale, there are many timescales used.[58] A single ruling time assumes an ideal clock as a reference. Even if one defers to such an ideal, it is through the mediation of material clocks, and such clocks break or function poorly. Awareness of this mundane fact generates business for clock shops whose records indicate that most of the watches and clocks they repaired they did not sell.[59] Moreover, that the difference between a sense of the ideal clock and the performance of one's watch/clock is a source of anxiety.[60] Time pluralism is a feature of the clock time being mediated by material objects. Another form of time pluralism is a feature of our existence in a universe with so many cycles and forces. This is not a social scientific observation, but one that the scientists who have argued about the leap second have come to recognize. As three scientists (an astrodynamics engineer and two astronomers) put it: "it is too much to expect to have one timescale for all purposes."[61] It is not just about duration and timing, but about the very way one organizes and represents one's own cycles and rhythms. Such conflicts are felt by those involved with time metrology who manage and design the global time system. To view this system as a stable monolith is false; to view it as being propelled into changes because of technological innovation and multiple political pressures is more accurate.

In addition to metrological challenges to uniform time standardization are the global political challenges. The leap second debate reveals these challenges because the deciding body, the World Radiocommunication Congress, had trouble reaching a consensus in its 2012 meeting. Such international institutions are sites where the universalizing logic of Western modernity meets global reluctance to agree to Western proposals. As Partha Chatterjee observes, "the postcolonial world outside of Western Europe and North American actually constitutes *most* of the populated modern world."[62] As the World Radiocommunication Congress demonstrates, the postcolonial world also controls most of the votes in such institutions—votes likely to passively resist through abstentions as has been the case with proposals to eliminate the leap second. The tension created by a universalizing perspective emerging out of a minority of the world's population makes the quest to impose and maintain a universal standard unstable in the long run.

Since the regulation and creation of time has such power over everyday life, it is a site of competition, and as Geoff Gordon points out, this competition often gets played out in law.[63] The establishing of the prime meridian was a momentous event in the history of international law, along with the establishment of uniform units of time.[64] Presently, Gordon notes, "there are only a select few global actors capable of deploying the expertise and resources, computational and infrastructural, necessary to negotiate temporalities and legal orders to their advantage internationally."[65] These developments do not represent an international consensus of equals, but technological and former colonial powers setting regulations for the entire global population. The northern hemispheric and Western bias of the institutions in charge of time exclude the vast majority of the world's population. It is also assumed that the basis of the positions in these debates are rationality and science rather than ideas privileged by a few cultural traditions. Occasionally nationalism breaks through, such as when a British minister complained that if the leap second was eliminated, the prime meridian for time would eventually be in New York.[66] Mostly, however, the localism and nationalisms are shrouded by claims of universalism.

Something Lost

The danger of multiple coexisting timescales that has been raised in debates about global times is not just that such timescales may exist, or that material clocks manifest these timescales, but that these multiple timescales would exist without many knowing it. This makes every clock potentially unknowingly unsettled. Technology so completely mediates between timekeeping and the use of clocks that many of those who rely on accurate timekeeping have no way of knowing the source of their time. The time has become a matter of faith. Even those who pay attention to the legal sources of time cannot distinguish an accurate from an inaccurate timestamp after the fact. Timestamps do not identify their source. In a world where a difference of nanoseconds is meaningful, indiscriminate faith in timestamps is a potential source of manipulation.

Maybe there is something that has been lost. Standardization eliminated diversity of rhythms; disciplining clocks disciplines us and elides alternative temporalities. The polyrhythms of daily life have been obscured by a single timescale. The special meaning of particular moments because of conjunctions and coincidences has disappeared in exchange for homogeneous, empty time. It is not that time is empty or homogeneous—only the chosen representation of time. Market cycles, the rhythms of the world studied by science, the physiological changes in bodies all continued, but instead of specialized timekeeping for each, there was a single timescale. Moreover, instead of timescales anchored to something observable, there emerged technoscientists who used algorithms and calculations to define and disseminate time. What we gained is the ability to relate time across space and to reliably measure duration—features critical to both bureaucracy and commerce. What the Guyanese writer Wilson Harris implies is that what has been lost is a connection to local time and rhythms. In contrast, when it comes to time, we are cast adrift from season and place and caught in a web of uniformity that extends everywhere there is a human presence.

Notes

1. Harris, "*Tradition, the Writer and Society*," 15.
2. European Space Agency, "Telling Time on the Moon."
3. Ravipati, "The Moon"; Dunn, "What Time Is It."
4. IEEE Standards Association, "Precision Clock Synchronization Protocol."
5. IEEE Standards Association, "Technical Terminology and Communications."
6. United States General Accounting Office, *Patriot Missile Defense.*
7. Weber, *Economy and Society*, 1,402.
8. Bergson, *Matter and Memory*, 67–68.
9. Felski, *Doing Time*, 16–17.
10. Glissant, *Caribbean Discourse*, 98.
11. Allen, *Republic in Time*, 9.
12. Postill, "Clock and Calendar Time," 251.
13. O'Connell, "Metrology," 130.
14. Birth, "Capital Flows"; Thompson, "Industrial Capitalism"; Postone, *Time, Labor and Social Domination.*
15. Stiegler's works that address this are: *What Makes Life Worth Living*, *Automatic Society*, *Age of Disruption*, and "For a Neganthropology."
16. Scott, *Seeing Like a State*, 349.
17. Vázquez, "Modernity, Coloniality, and Visibility," 2.2.
18. Griffiths, *A Sideways Look at Time*, 17.
19. see Birth, "Non-clocklike Features"; Birth, "Time Blind," chapter 2; Schmidgen, "Brain-Time Experiments."
20. Deane, "Instruments and Observation"; Jami, "Western Devices"; Sivin, *Granting the Seasons*; Sôma et al., "Units of Time"; Wu "The Hong Kong Clock."
21. Edmonson, *Book of the Year*; Rice, *Maya Political Science*, and *Maya Calendar Origins.*
22. Wishnitzer, *Reading Clocks.*
23. Rooney, *About Time*, chapter 6.
24. Wu, "The Hong Kong Clock," 339.
25. Rice, *Maya Political Science*, 56.
26. Wishnitzer, *Reading Clocks*; "At Approximately Eleven."
27. Smith, *Wealth of Nations*, 36.
28. Strathern, *Reproducing the Future*, 17.
29. Conrad, "Nothing is the Way it Should Be," 845–7.
30. Chertisch et al., *Anthropologies of Revolution*, 150.
31. Swinehart, "Decolonial Time," 101.
32. Swinehart, "Decolonial Time," 103.
33. Crary, *24/7*, 9.
34. Stiegler, *Automatic Society*, 109–10.
35. Baird, *Thing Knowledge*, 4.
36. Canales, *Tenth of a Second.*
37. Baird, *Thing Knowledge*, 10.
38. Bartky, *One Time Fits All.*
39. Vico, *The New Science*, 138; Berlin, *Vico and Herder*, 100–5.
40. Berlin, *Vico and Herder*, 13.
41. Attali, *Histoires du temps*, 32.
42. Austin, *How to do Things with Words.*

43. Panfilo et al., "First Step"; Matsakis, "Generalizable Formalism."
44. Heidegger, *The Question Concerning Technology.*
45. Henzi et al., "Patterns of Movement"; see also Dunbar et al. "Time as an Ecological Constraint."
46. Bennie et al., "Biogeography of Time Partitioning."
47. Leroi-Gourhan, *Gesture and Speech,* 349 and 359.
48. Bishop and Ross, "Technics," 115.
49. Adam, "Un/Learning."
50. Benjamin, "Philosophy of History," 261.
51. Aristotle, *Physics,* IV.14.
52. See Serres and Latour, *Conversations on Science,* 60.
53. Deleuze, "Societies of Control."
54. Foucault, *Discipline and Punish.*
55. Green, "On the Move"; Keightley, "From Immediacy to Intermediacy"; Pschetz, "Time Affordances."
56. Ortiz, *Cuban Counterpoint.*
57. Deleuze and Guattari, *A Thousand Plateaus,* 21–22.
58. Rooney, *About Time,* 76.
59. McCrossen, "Pocket Watches," 14, and "Broken Watches," 209.
60. McCrossen, "Broken Watches," 209.
61. Seago et al., "Requirements for UTC," 5.
62. Chatterjee, "Anderson's Utopia," 132.
63. Gordon, "Imperial Standard Time," 1,198.
64. Gordon, "Imperial Standard Time," 1,202 and 1,204; see also International Meridian Conference, *International Conference*; Barrows, *Cosmic Time of Empire.*
65. Gordon, "Imperial Standard Time," 1,221.
66. Swineford, "Greenwich Mean Time."

Glossary

Note: for many metrological institutions, the convention is to base the initials off of the French name even when the institution is referred to with the English name. Both are given here.

Accuracy	The ability to determine the correct time. This is different from precision. A clock can indicate the correct time and still not represent small increments of time.
Astronomical Hours	A count of twenty-four hours of equal length that begins when the Sun crosses the meridian.
Babylonian Hours	A timescale of twenty-four hours of equal length that begins at sunrise.
BIH	Bureau international de l'heure/International Bureau of the Hour. Its duties defining UTC were taken over by the BIPM and its duties charting Earth rotation were taken over by IERS.
Bohemian Hours	See Italian Hours.
BIPM	Bureau international des poids et mesures/ International Bureau of Weights and Measures. This is the agency that carries out CGPM policies.
Canonical Hours	These are hours that divide the daylight period into twelve equal segments. As a result, they vary with the season and with latitude.
CCIR	Comité consultif international pour la radio/ Consultative Committee on International Radio, the predecessor of the ITU-R.
CCTF	Consultative Committee on Time and Frequency.

CGPM	Conférence Générale des poids et mesures/General Conference on Weights and Measures¾an intergovernmental organization that sets policy for the BIPM.
CIPM	Comité international des poids et mesures/International Committee for Weights and Measures—a group of eighteen scientists appointed by the CGPM to make recommendations to the CGPM.
EAL	Echelle atomique libre/Free Atomic Scale, a timescale used to determine TAI.
ERA	Earth Rotation Angle. This is a measure of Earth's rotation relative to a fixed location on the celestial equator.
GNSS	Global Navigation Satellite System.
French Hours	See Small Hours.
Greek Hours	In Renaissance and Early Modern Europe, this term referred to a timescale consisting of twenty-four hours of equal length beginning at midnight.
IEEE	Institute of Electrical and Electronics Engineers, the body which determines the POSIX standards.
IERS	International Earth Rotation and Reference Systems Service. Formerly part of the BIH, this institution charts Earth Rotation relative to UTC and makes the recommendation for when a leap second is to take place.
Italian Hours	This is a count of twenty-four equal hours that begins at the ringing of the Angelus about a half hour after sunset.
ITU-R	Radiocommunication Sector of the International Telecommunications Union. This is a treaty-based organization.
Mean Solar Time	This is a timescale based on a day length that averages the length of day throughout the year at a specified longitude. Because of the Earth's tilt relative to the Sun, its elliptical orbit, and because as the Earth moves a 360-degree rotation does not bring it back to the same orientation to the Sun, the Earth's day varies slightly in length. Most clocks

	represent a mean time. Greenwich Mean Time is the mean time calculated for the meridian of Greenwich, UK.
NIST	United States National Institute of Standards and Technology
ns	Nanoseconds—a nanosecond is one billionth of a second.
NTP	Network Time Protocol—a protocol for synchronizing computer networks with UTC.
POSIX	The computer protocol standards that allow for compatibility between different operating systems.
Precision	The ability to measure small increments reliably. This is different from accuracy. A clock can measure tiny increments of time reliably but still indicate the incorrect time if it is not set properly.
PTB	Physikalisch-Technische Bundesanstalt/The National Metrology Institute of Germany.
PTP	Precision Time Protocol—a protocol for synchronizing computer networks to a simulacrum of TAI produced from receiving time signals from GPS satellites.
SI	Système Internationale—the international system of metrological standards.
Sidereal Time	This is time measured by the rotation of stars around the celestial pole. The sidereal day is 23 hours, 56 minutes, and 4.1 seconds long.
Small Hours	This is a system of two series of twelve equal hours with one series beginning at midnight and the other at noon. This is the most common form of counting hours today.
TAI	Temps atomique international/International Atomic Time. A coordinate timescale that is a continuous count of seconds.
Temporal Hours	see Canonical Hours.
USNO	United States Naval Observatory, the source of atomic time for GPS
UTC	Coordinated Universal Time—this abbreviation is not based on either the French or English, but on a variant of the abbreviation UT (universal time).

UTC(k)	This is a locally created version of UTC.
UTC(USNO)	This is the version of UTC created by the USNO.
UT	Universal Time.
UT1	A mean time that represents the observed rotation of Earth corrected for polar motion.

Bibliography

Adam, Barbara. *Timewatch: The Social Analysis of Time*. Oxford: Polity Press, 1995.

———. *Timescapes of Modernity*. London: Routledge, 1998.

———. "Un/Learning the Habits of Clock Time." *Counterpoints* 62 (1999): 103–28.

———. *Time*. Cambridge: Polity, 2004.

Achkar, Joseph. "Progress on Time Signals and Frequency Standard Emissions." *ITU News Magazine*, no. 20 (2023): 12–14.

Addyman, Peter. *York. Historic Towns Atlas, Volume V.* Oxford: Historic Towns Trust, 2015.

Adelard of Bath. *Conversations with his Nephew: On the Same and the Different, Questions on Natural Science, and On Birds*, translated by Charles Burnett. Cambridge: Cambridge University Press, 1998.

Adib, and Paul Emiljanowicz. "Colonial Time in Tension." *Time and Society* 28, no. 3 (2017): 1,221–38.

Aethelwold. *Regularis Concordia Anglicae Nationis Monachorum Sanctimonialiumque* [The Monastic Agreement of the Monks and Nuns of the English Nation], translated by Dom Thomas Symons. London: Thomas Nelson and Sons, 1953.

Akerman, James R. "Introduction." In *The Imperial Map: Cartography and the Mastery of Empire*, edited by James R. Akerman, 1–9. Chicago: University of Chicago Press, 2009.

Alcuin. "de Disputatio Puerorum" [The Disputation of Boys]. In *Patrologia Latina, volume 101*, edited by J.-P. Migne, columns 1,097–144. Turnhout: Typographi Brepols Editores Pontifici, 1844–1864.

Alexander, DeAlva Stanwood. *A Political History of the State of New York, volume II, 1833–1861*. Port Washington, NY: Ira J. Friedman, 1969.

Alfonso. *Tabule Astronomice* [Astronomical Tables]. Venice: Johannis Hamman, 1492.

———. A Transcription of the *Libro de las tablas alfonsies* [Book of the Alfonsine Tables]. In *The Alfonsine Tables of Toledo*, edited by José Chabás and Bernard R. Goldstein, 9–94. Dordrecht: Springer Science, 2003.

Allan, David W., B. E. Blair, D. D. Davis, and H. E. Machlan. "Precision and Accuracy of Remote Synchronization via Network Television Broadcasts, Loran-C, and Portable Clocks." *Metrologia* 8, no. 2 (1972): 64–72.

Allen, S. L. "Telescope Systems at Lick Observatory and Keck Observatory." In *Decoupling Civil Timekeeping from Earth Rotation.* American Astronautical Society, Science and Technology Series, vol. 113, 289–96. San Diego: Univelt, 2011.

———. "Planes Will Crash! Things that Leap Seconds Didn't, and Did, Cause." In *Requirements for UTC and Civil Timekeeping on Earth*, edited by J. H Seago, R. L.

Seaman, P. K. Seidelmann, and S. L. Allen, 15–26. American Astronautical Society, Science and Technology Series, vol. 115. San Diego: Univelt, 2013.

Allen, Thomas M. *A Republic in Time: Temporality and Social Imagination in Nineteenth-Century America.* Chapel Hill: UNC Press, 2008.

Allison, K. J. "Markets and Fairs." In *A History of Yorkshire: The City of York*, edited by P. M. Tillott, 484–91. London: Oxford University Press, 1961.

Alsted, Johann Heinrich. *Encyclopaedia Septem tomis distincta* [An Encyclopedia in Seven Distinct Volumes]. Herborn: G. Corvini, 1630.

Ammann, Jost. *Vergleichung nachuolgender Siben Uhren* [Comparison of the Following Seven Clocks]. Germanische Nationalmuseum. Inventory number HB2698, 1589. Retrieved 5 January 2026 from https://objektkatalog.gnm.de/wisski/navigate/23464/view.

Anderson, Benedict. *Imagined Communities*, new edn. London: Verso, 2006.

Angel, James. "When Finance Meets Physics: The Impact of the Speed of Light on Financial Markets and Their Regulation." *Financial Review* 49, no. 2 (2014): 271–81.

Anonymous. *Here endith a compendiouse treetise dyalogue. of diues [and] paup[er].* London: Richard Pynson, 1493.

Anonymous. "Licensing Act: Important Conference of Licensed Victuallers at Birmingham." *Brewer's Guardian* 3, 14 January 1873: 3–7.

Apian, Petrus (with notes by Wilhelm IV and Andreas Schöner). *Astronomicum Caesareum* [The Astronomy of the Caesars]. Ingolstadt: Peter Apian, 1540.

Appadurai, Arjun. "Introduction: Commodities and the Politics of Value." In *The Social Life of Things: Commodities in the Cultural Perspective*, edited by Arjun Appadurai, 3–63. Cambridge: Cambridge University Press, 1986.

Arias, Elisa Felicitas. "The Metrology of Time." *Philosophical Transactions of the Royal Society A: Mathematical, Physical and Engineering Sciences* 363, no. 1834 (2005): 2,289–305.

———. "Perspective: Time Scales and Clocks." *Review of Scientific Instruments* 83, no. 2 (2012). https://doi.org/10.1063/1.3681927.

———. "One More Second." *Nature Physics*, no. 12 (2016): 1,178.

Arias, E. F., G. Panfilo, and G. Petit. "Timescales at the BIPM." *Metrologia* 48, no. 4 (2011): S145–53.

Arias, E. F., and G. Petit. "The Hyperfine Transition for the Definition of the Second." *Annalen der Physik* [Annals of Physics] 531, no. 5 (2019): 1–12.

Arias, E. Felicitas, Demetrios Matsakis, Terry J. Quinn, and Patrizia Tavella. The 50th Anniversary of the Atomic Second. *IEEE Transactions on Ultrasonics, Ferroelectrics, and Frequency Control* 65, no. 6 (2018): 898–903.

Aristotle. *Physics, Volume 1: Books I–IV*, translated by P. H. Wicksteed and F. M. Cornford. Loeb Classical Library 228, 1957.

Arnold, John H., and Caroline Goodson. "Resounding Community: The History and Meaning of Medieval Church Bells." *Viator* 43, no. 1 (2012): 99–130.

Ashworth, William. "The Habsburg Circle." In *Patronage and Institutions: Science, Technology, and Medicine at the European Court*, edited by Bruce T. Moran, 137–68. Rochester, NY: Boydell, 1991.

Atkinson, Niall. "Sonic Armatures: Constructing an Acoustic Regime in Renaissance Florence." *Senses and Society* 7, no. 1 (2012): 39–52.

———. "The Republic of Sound: Listening to Florence at the Threshold of the Renaissance." *I Tatti Studies in the Italian Renaissance* 16, no. 1–2 (2013): 57–84.

———. *The Noisy Renaissance: Sound, Architecture, and Florentine Urban Life.* University Park, PA: Pennsylvania State University Press, 2016.

Atlanta Constitution. "Groover Will Buy New House Clock." 24 February 1964: 1 and 6.

Attali, Jacques. *Histoires du temps.* Paris: Fayard, 1982.

Audoin, Claude, and Bernard Guinot. *The Measurement of Time: Time Frequency and the Atomic Clock*, translated by Stephen Lyle. Cambridge: Cambridge University Press, 2001.

Austin, J. L. *How to Do Things with Words*. Cambridge: Harvard University Press, 1962.

Baader, J. "Wann man vor 400 Jahren zu Nürnberg zu und von der Arbeit ging" [When people went to and from work 400 years ago]. *Zeitschrift für Kulturgeschichte* [Journal of Cultural History], no. 3 (1858): 229–31.

Bailey, F. G. *Humbuggery and Manipulation: The Art of Leadership*. Ithaca, NY: Cornell University Press, 1988.

Baird, Davis. *Thing Knowledge: A Philosophy of Scientific Instruments*. Berkeley: University of California Press, 2004.

Banerjee, Parameswar, and Matsakis, Demetrios. *An Introduction to Modern Timekeeping and Time Transfer*. Cham, Switzerland: Springer Nature, 2023.

Barak, On. *On Time: Technology and Temporality in Modern Egypt*. Berkeley: University of California Press, 2013.

Barber, Richard. "The Norman Conquest and the Media." *Anglo-Norman Studies* XXVI, edited by John Gillingham, 1–21. Woodbridge: The Boydell Press, 2004.

Barnes, J. A., and G. M. R. Winkler. *The Standards of Time and Frequency in the USA*. Washington, DC: US Government Printing Office, 1974.

Barrows, Adam. *The Cosmic Time of Empire*. Berkeley: University of California Press, 2011.

Barry, Andrew. *Political Machines*. London: Althone, 2001.

Bartholomaeus Anglicus. *De genuinis rerum coelestium, terrestrium et infernarum proprietatibus libri XVIII* [On the True Properties of Celestial, Terrestrial, and Infernal Things]. Frankfurt: Wolfgang Richter, 1601.

Bartky, Ian. "The Adoption of Standard Time." *Technology and Culture* 25, no. 1 (1989): 25–56.

———. *Selling the True Time: Nineteenth-Century Timekeeping in America*. Stanford: Stanford University Press, 2000.

———. *One Time Fits All*. Stanford: Stanford University Press, 2007.

Basing, Patricia. *Trades and Crafts in Medieval Manuscripts*. London: British Library, 1990.

Bastian, Michelle. "Liberating Clocks: Exploring the Transformative Potential of 'Clock-Time.'" *New Formations: A Journal of Culture/Theory/Politics* 92, no. 1 (2017): 41–55.

Battersby, Christine. *The Phenomenal Woman: Feminist Metaphysics and the Patterns of Identity*. Cambridge: Polity Press, 1998.

Bear, Laura. "Time as Technique." *Annual Review in Anthropology* 45 (2016): 487–502.

———. "Anthropological Futures." *Social Anthropology* 25, no. 2 (2017): 142–58.

Beard, Ronald L. "Role of the ITU-R in Time Scale Definition and Dissemination." *Metrologia* 48, no. 4 (2011): S125–31.

Beaver, Donald deB. "Bernard Walther: Innovator in Astronomical Observation." *Journal of the History of Astronomy* 1, no. 1 (1970): 39–43.

Bede. *Ecclesiastical History*, translated by J. E. King. Cambridge, MA: Loeb Classical Library, 1930.

———. *The Reckoning of Time*, translated by Faith Wallis. Liverpool: Liverpool University Press, 1999.

Bedini, Silvio. "The Evolution of Science Museums." *Technology and Culture* 6, no. 1 (1965): 1–29.

———. *Patrons, Artisans and Instruments of Science, 1600–1750*. Aldershot: Ashgate Variorum, 1999.

Beecher, Catherine. *A Treatise on Domestic Economy*. New York: Harper and Brothers, 1845.

Beehler R. "Role of the Consultative Committee on International Radio (CCIR) in Time and Frequency." In *Proceedings of the 23rd Annual Precise Time and Time Interval*

(PTTI) Applications and Planning Meeting (Pasadena, CA), 321–29. NASA Conference Publication 3159, 1992.

Bendefy, Lázló. "Regiomontanus und Ungarn [Regiomontanus and Hungary]." In *Regiomontaus-Studien* [Regiomontanus Studies], edited by Günther Hamann, 243–53. Vienna: Verlag der Österreichischen Akademie der Wissenschaften, 1980.

Benjamin, Walter. "Theses on the Philosophy of History." In *Illuminations: Essays and Reflections*, edited by Hannah Arendt, 253–64. New York: Schocken, 1968.

Benedict of Nursia. *Rule of St. Benedict*, translated by Timothy Frye. New York: Vintage, 1998.

Bennett, Jane. *Vibrant Matter: A Political Ecology of Things*. Durham, NC: Duke University Press, 2010.

Bennie, Jonathan J., James P. Duffy, Richard Inger, and Kevin J. Gaston. "Biogeography of Time Partitioning in Mammals." *Proceedings of the National Academy of Sciences* 111, no. 38 (2014): 13,727–32.

Benson, George. *York Bellfounders*. Yorkshire Philosophical Society Annual Report for 1898. 17–35.

Berggren, J. Lennart, and Alexander Jones. "Introduction." In *Ptolemy's Geography: An Annotated Translation of the Theoretical Chapters*, 3–54. Princeton: Princeton University Press, 2000.

Bergson, Henri. *Matter and Memory*, translated by N. M. Paul and W. S. Palmer. New York: Zone Books, 1988.

———. *Duration and Simultaneity*, translated by Mark Lewis and Robin Durie. Manchester: Clinamen Press, 1999.

Bergström, John A. "A Type of Pendulum Chronoscope and Attention Apparatus." *Psychological Review* 7, no. 5 (1900): 483–88.

Berlin, Isaiah. *Vico and Herder: Two Studies in the History of Ideas*. New York: Viking, 1976.

Berthoud, Ferdinand. 1802. *Histoire de la mesure du temps par les horloges* [History of the Measure of Time by Clocks]. 2 vols. Paris: Imprimerie de la République, 1802.

Bhabha, Homi. *The Location of Culture*. New York: Routledge, 1994.

Biagioli, Mario. "Galileo's System of Patronage." *History of Science* 28 (1990): 1–62.

BIH. *Bulletin Horaire*. May–June, no. 3, 1955.

Bilfinger, Gustav. *Der Bürgerliche Tag* [The Civil Day]. Stuttgart: W. Kohlhammer, 1988.

———. *Mittelalterlichen Horen und die Modernen Stunden* [Medieval Hora and Modern Hours]. Stuttgart: W. Kohlhammer, 1892.

BIPM. *Consultative Committee for Time and Frequency (CCTF) 14th Meeting*, April 20–22. Sèvres: BIPM, 1999.

BIPM Time Section. *Annual Report of the BIPM Time Section*. Vol. 1. Sèvres: BIPM, 1988.

BIPM/ITU. "Memorandum of Understanding." BIPM website, 2020. Retrieved 5 January 2026 from https://www.bipm.org/en/liaison-partners/itu.

Birmingham Daily Post. "United Towns' Association of Licensed Victuallers Annual Meeting." 21 August 1873: 6.

———. "Important to Publicans." 18 November 1874: 5.

———. "Aston Petty Sessions: Friday." 16 April 1877: 6.

Birth, Kevin. *Any Time is Trinidad Time: Social Meanings and Temporal Consciousness*. Gainesville, FL: University Press of Florida, 1999.

———. "The Immanent Past: Culture and Psyche at the Juncture of Memory and History. Introduction to the special issue 'The Immanent Past.'" *Ethos* 34, no. 2 (2006): 169–91.

———. "Past Times: Temporal Structure of History and Memory." *Ethos* 34, no. 2 (2006): 192–210.

———. "Time and the Biological Consequences of Globalization." *Current Anthropology* 48, no. 2 (2007): 215–36.

———. *Objects of Time: How Things Shape Temporality*. New York: Palgrave, 2012.

———. Breguet's Decimal Clock: A Masterpiece from the Enlightenment. *Frick Collection Member's Magazine* Winter (2014): 10–11.

———. Non-clocklike Features of Psychological Timing and Alternatives to the Clock Metaphor. *Timing and Time Perception* 2, no. 3 (2014): 312–24.

———. *Time Blind: Problems in Perceiving Other Temporalities*. New York: Palgrave Macmillan, 2016.

———. "While the West Sleeps: Deglobed Globalization and Its Consequences". In *Time, Globalization and Human Experience*, edited by Paul Huebener, Susie O'Brien, Tony Porter, Liam Stockdale, and Yanqiu Rachel Zhou, 109–26. London: Routledge, 2017.

———. "Time Lords, Standards and Rhizomatic Imperialism." In *Ethnographies of U.S. Empire*, edited by Carole McGranahan and John Collins, 227–45. Durham, NC: Duke University Press, 2018.

———. "Standards in the Shadows for Everyone to See: The Supranational Regulation of Time and the Concern over Temporal Pluralism." In Law *and Time*, edited by Siân M Beynon-Jones and Emily Grabham, 196–211. London: Routledge, 2019.

———. 2022. "Capital Flows, Itinerant Laborers, and Time." *Time and Society* 30, no. 3 (2022): 392–414.

Bishop, Ryan, and Daniel Ross. "Technics, Time, and the Internation." *Theory, Culture and Society* 38, no. 4 (2021): 111–33.

Bize, Sébastien. "The Unit of Time: Present and Future Directions." *Comptes Rendus Physique* [Physics Reports] 20, nos. 1–2 (2019): 153–68.

Black, Jeremy. *Maps and Politics*. London: Reaktion Books, 1997.

Bloch, Marc. *Feudal Society*. Chicago: University of Chicago Press, 1961.

Blondé, Bruno, and Gerrit Verhoevent. "Against the Clock." *Continuity and Change* 28, no. 2 (2013): 213–44.

Bluedorn, Allan C. *The Human Organization of Time: Temporal Realities and Experience*. Stanford: Stanford University Press, 2002.

Bobinger, Maximilian. *Kunstuhrmacher in Alt-Augsburg* [Artistic Clockmakers in Old Augsburg]. Augsburg: Hans Rosler Verlag, 1969.

Boer, Bas de, Hedwig te Molder, and Peter-Paul Verbeek. "The Perspective of the Instruments: Mediating Collectivity." *Foundations of Science* 23 (2018): 739–55.

Boethius. *Fundamentals of Music*, translated by Calvin M. Bower, edited by Claude V. Palisca. New Haven: Yale University Press, 1989.

Bonham, Milledge L. "New York and the Election of 1860." *New York History* 15, no. 2 (1934): 124–43.

Bonhoeffer, Dietrich. *Creation and Fall*. New York: Macmillan, 1959.

Bourdieu, Pierre. *On the State: Lectures at the Collège de France, 1989–1992*, translated by David Fernbach, edited by Patrick Champagne, Remi Lenoir, Franck Poupeau, and Marie-Christine Rivière. Cambridge: Polity Press, 2014.

Bradbury, Nancy Mason, and Carolyn P. Collette. "Changing Time: The Mechanical Clock in Late Medieval Literature." *The Chaucer Review* 43, no. 4 (2009): 351–75.

Brahe, Tycho. *Tychonis Brahe astronomiae instauratae mechanica* [Tycho Brahe's instruments for the restoration of astronomy]. Wandesburg: n.p., 1598.

———. *Tychonis Brahe astronomiae instauratae mechanica* [Tycho Brahe's instruments for the restoration of astronomy]. Nuremberg: Levinus Hulsius, 1602.

Brand, Paul. "Lawyers' Time in England in the Later Middle Ages." In *Time in the Medieval World*, edited by Chris Humphrey and W. M. Omrod, 73–104. York: York Medieval Press, 2001.

Brandi, Karl. *The Emperor Charles V*, translated by C. V. Wedgwood. London: Jonathan Cape, 1939.

Brooklyn Evening Star. "The Harmonious Democracy." Vol. 50, no. 5766, 15 September 1859: 2.

Brown, Bill. "Thing Theory." *Critical Inquiry* 28, no. 1 (2001): 1–22.

Buffalo Commercial. "Democratic State Convention." Vol. 25, no. 8737, 15 September 1859: 1.

Buffalo Daily Courier. "Latest News By Telegraph: Democratic State Convention." No. 194, 15 September 1859: 3.

———. "Fast Time." No. 200, 22 September 1859: 2.

Bukovinská, Beket. "The *Kunstkammer* of Rudolf II: Where it Was and What it Looked Like." In *Rudolf II and Prague: The Court and the City*, edited by Eliška Fučiková, James M. Bradburne, Beket Bukovinská, Jaroslava Hausenblasová, Lumomir Konečy, Ivan Muchka, and Michal Šroněk, 199–208. London: Thames and Hudson, 1997.

Burton, Eric. *The History of Clocks and Watches.* New York: Crescent, 1979.

Burke, Peter. "Reflections on the Cultural History of Time." *Viator* 35 (2004): 617–26.

Burton, Janet, trans. *The Foundation History of the Abbeys of Byland and Jervaulx.* York: University of York, 2006.

Callon, Michel, and Bruno Latour. "Unscrewing the Big Leviathan." In *Advances in Social Theory and Methodology*, edited by K. Knorr-Cetina and A. Cicourel, 277–303. London: Routledge and Kegan Paul, 1981.

Canales, Jimena. *A Tenth of a Second.* Chicago: University of Chicago Press, 2009.

———. *The Physicist and the Philosopher.* Princeton: Princeton University Press, 2015.

Cassini, Giovanni Dominico. *Recueil d'observations faites en plusieurs voyages par ordre de sa majesté pour perfectionner l'astronomie et la géographie avec divers traitez astronomiques* [Collection of observations made on several voyages by order of his majesty in order to improve astronomy and geography with various astronomical treatises]. Paris: L'Imprimerie Royale, 1693.

Castells, Manuel. *The Rise of the Network Society.* Oxford: Blackwell, 2000.

CCIR. *XII Plenary Assembly, Volume III.* Geneva: International Telecommunications Union, 1970.

———. *XII Plenary Assembly, Volume VII.* Geneva: International Telecommunications Union, 1970.

CCDS. *Procès-Verbaux de Séances.* Paris: International Bureau of Weights and Measures, 1958.

———. *3rd Session.* Paris: International Bureau of Weights and Measures, 1963.

———. *5th Session.* Paris: International Bureau of Weights and Measures, 1970.

CCTF. CCTF Strategy Document, May 2016, covering 2016–2026.

CGPM. *Comptes Rendus des Séances de la Dixième Conférence Génerale des Poids et Mesures* [Proceedings of the tenth general conference on weights and measures]. Sèvres: BIPM, 1954.

CGPM. "Resolution 1: SI Unit of Time (second)." *Comptes rendus des séances de la treizième conférence générale des poids et measures* [Proceedings of the thirteenth general conference on weights and measures]. Sèvres: BIPM, 1967/8.

CGPM 2018. "Resolution 1: On the Revision of the International System of Units (SI)." *Proceedings of the 26th Meeting of the General Conference on Weights and Measures*, 2018: 472–74.

CGPM 2022. "Resolution 4: On the Use and Future Development of UTC," *27th Meeting of the General Conference on Weights and Measures*, 2022. Retrieved 11 January 2023 from https://www.bipm.org/en/cgpm-2022/resolution-4.

Chabás, José, and Bernard Goldstein. "Introduction." In *The Alfonsine Tables of Toledo*, 1–8. Dordrecht: Spring Science, 2003.

———. *A Survey of European Astronomical Tables in the Late Middle Ages.* Leiden: Brill, 2012.

Chadsey, H., and D. McCarthy. "Relating Time to the Earth's Variable Rotation." *32nd Annual Precise Time and Time Interval (PTTI) Meeting* (2001): 237–54.

Chakrabarty, Dipesh. *Provincializing Europe: Postcolonial Thought and Historical Difference.* Princeton: Princeton University Press, 2000.

Chandler, Bruce, and Clare Vincent. "To Finance a Clock: An Example of Patronage in the 16th Century." In *Clockwork Universe: German Clocks and Automata, 1550–1650*, edited by Klaus Maurice and Otto Mayr, 103–13. New York: N. Watson Academic Publications, 1980.

Chapman, Allan. "Tycho Brahe—Instrument Designer, Observer and Mechanician." *Journal of the British Astronomical Association.* 99, no. 2 (1989): 70–76.

Charvin, Pierre. "Allocutions." In *75 ans au service de l'heure universelle.* Paris: Observatoire de Paris, Bureau des longitudes, 1992.

Chatterjee, Partha. "The Nationalist Resolution of the Women's Question." In *Recasting Women: Essays in Indian Colonial History*, edited by Kumkum Sangara and Sudesh Vaid, 233–53. New Brunswick, NJ: Rutgers University Press, 1990.

———. "Anderson's Utopia". *Diacritics* 29, no. 4 (1999): 128–34.

Chi, A. R., and H. S. Fosque. "A Step in Time: Changes in Standard Frequency and Time Signal Broadcasts." *IEEE Spectrum* 9 (1972): 82–86.

Chertisch, Igor, Martin Holbraad, and Nico Tassi. *Anthropologies of Revolution.* Berkeley: University of California Press, 2020.

CIPM. *Procés-verbaux des séances, session de 1956* [Report of the 1956 Meetings]. Paris: Gauthier-Villars.

CIPM. *58th Session.* Paris: International Bureau of Weights and Measures, 1969.

Cipolla, Carlo M. *Clocks and Culture 1300–1700.* New York: Norton, 1978.

Clark, Andy. *Microcognition: Philosophy, Cognitive Science and Parallel Distributive Processing.* Cambridge, MA: MIT Press, 1989.

———. *Supersizing the Mind: Embodiment, Action, and Cognitive Extension.* Oxford: Oxford University Press, 2008.

Clark, Andy, and David Chalmers. "The Extended Mind." *Analysis* 58, no. 1 (1998): 7–19.

Cleempoel, Koenraad van. "Philip II's Escorial and its Scientific Instruments." In European *Collections of Scientific Instruments, 1550–1750*, edited by Giorgio Strano, Stephen Johnston, Mara Miniati, and Alison Morrison-Low, 101–27. Leiden: Brill, 2009.

Clemence, Gerald M. "On the System of Astronomical Constants." *The Astronomical Journal.* 53, no. 6 (1948): 169–79.

———. "On Revising the Official System of Astronomical Constants." In *Constantes Fondamentales de l'Astronomie* [Astronomy's Fundamental Constants], 19–28. Paris: Centre National de la Recherche Scientifique, 1950.

Colloque International de Chronometrie [International Conference of Chronometry]. "Introduction." 16–20 September 1969: 1–9.

Colvin, H. M., ed. *Building Accounts of King Henry III.* Oxford: Clarendon Press, 1971.

Comitas, Lambros. "Occupational Multiplicity in Rural Jamaica." In *Work and Family Life: West Indian Perspectives*, edited by Lambros Comitas and David Lowenthal, 156–73. Garden City, NY: Anchor, 1973.

Connable, Alfred, and Ed Silberfarb. *Tigers of Tammany: Nine Men Who Ran New York.* New York: Holt, Rinehart, and Winston, 1967.

Conrad, Sebastian. "Nothing is the Way It Should Be." *Modern Intellectual History* 15, no. 3 (2018): 821–48.

Corbin, Alain. *Village Bells: Sound and Meaning in the Nineteenth-century French Countryside*, translated by Martin Thom. New York: Macmillan, 1998.

Crary, Jonathan. *24/7: Late Capitalism and the Ends of Sleep.* London: Verso, 2014.

Craster, H. H. E., and M. E. Thornton. *The Chronicle of St. Mary's Abbey, York: From Bodley MS. 39.* Surtees Society, CXLVIII. Durham: Andrews and Co., 1934.

Crease, Robert. Charles Sanders Peirce and the First Absolute Measurement Standard. *Physics Today* 62, no. 12 (2009): 39–44.

Crosby, Alfred W. *The Measure of Reality: Quantification and Western Society, 1250–1600.* Cambridge: Cambridge University Press, 1997.

Cuming, Syer. "Proceedings of the Association." *Journal of the British Archaeological* Association 4 (1849): 133–41.

Curtze, M. "Der Briefwechsel Regiomontan's mit Giovanni Bianchini, Jacob von Speier und Christian Roder." [Regiomontanus' correspondence with Giovanni Bianchini, Jacob von Speier, and Christian Roder]. *Abhandlungen zur Geschichte der mathematischen Wissenschaften mit Einschluss inhrer Anwendungen*, 12 (1902): 185–336.

Dackerman, Susan. *Prints and the Pursuit of Knowledge in Early Modern Europe.* New Haven: Yale University Press, 2011.

Daily Evening Bulletin. 1883. "The Change of Time: How the Clocks Will Be Changed." 17 November 1883: 1.

Danjon, A. "Le Temps: sa Définition Pratique, sa Mesure" [Time: Its Practical Definition, Its Measure]. *L'Astronomie* [Astronomy] 43 (1929): 13–23.

———. "La Rotation de la Terre" [The Rotation of the Earth]. *L'Astronomie* [Astronomy] 43 (1929): 115–28.

Davies, Arthur. 1977. Behaim, Martellus and Columbus. *The Geographical Journal* 143, no. 3 (1977): 451–59.

Davis, James. *Medieval Market Morality: Life, Law and Ethics in the English Marketplace, 1200–1500.* Cambridge: Cambridge University Press, 2012.

Davison, Graham. "Punctuality and Progress: The Foundations of Australian Standard-Time.*" Australian Historical Studies* 25, no. 99 (1992): 169–91.

Deane, Thatcher E. "Instruments and Observation at the Imperial Astronomical Bureau during the Ming Dynasty." *Osiris* 9, no. 1 (1994): 126–40.

Deines, Steven D., and Carol A Williams. "Earth's Rotational Deceleration: Determination of Tidal Friction Independent of Timescales." *The Astronomical Journal* 151, no. 4 (2016). Doi:10.3847/0004-6256/151/4/103.

Deleuze, Gilles. 1991. *Bergsonism.* New York: Zone Books, 1991.

———. "Postscript on the Societies of Control." *October* 59, winter (1992): 3–7.

Deleuze, Gilles, and Félix Guattari. *A Thousand Plateaus: Capitalism and Schizophrenia*, translated by Brian Massumi. Minneapolis: University of Minnesota Press, 1989.

Democratic Party. *Official Proceedings of the Democratic National Convention Held in 1860, at Charleston and Baltimore: Proceedings at Charleston.* Cleveland: Nevins, 1860.

Diffie, Bailey W., and George D. Winius, *Foundations of the Portuguese Empire, 1415–1580.* Minneapolis: University of Minnesota Press, 1977.

Dirklik, Arif. "How the Grinch Hijacked Radicalism." *Postcolonial Studies* 2, no. 2 (1999): 149–63.

Dobner, Gelasius. *Monumenta historica Boemiae, Tomus III* [Historical Monuments of Bohemia, Volume 3]. Prague: Clauser, 1774.

Dohrn-van Rossum, Gerhard. "The Diffusion of the Public Clocks in the Cities of late Medieval Europe 1300–1500." In *La Ville et l'Innovation: relais et réseaux de diffusion en Europe 14e - 19e siècles* [The City and Innovation: Relays and Diffusion Networks in Europe, 14th-19th Centuries], edited by B. Lepitit and J. Hoock, 29–43. Paris: Éditions de l'école des hautes etudes en sciences sociales, 1987.

———. "Zeit der Kirche—Zeit der Händler—Zeit der Städte" [Time of the Church—Time of the Merchants—Time of the Cities]. In *Zerstörung und Wiederaneignung von Zeit* [Destruction and Reappropriation of Time], edited by Rainer Zoll, 89–188. Frankfurt am Main: Suhrkamp, 1988.

———. "Schaguhr und Zeitorganisation: Zur frühen Geschichte der öffentlichen Uhren und den sozialen Folgen der modernen Stundenrechnung" [The Striking Clock and Time Organization: On the Early History of Public Clocks and the Social Consequences of Modern Hour Reckoning]. In *Im Netz der Zeit* [In the Web of Time], edited by Rudolf Wendorff, 49–60. Stuttgart: S. Hirzel, 1989.

———. *History of the Hour*, translated by Thomas Dunlap. Chicago: University of Chicago Press, 1996.

———. 2005. "Uhren, Glocken und Zeitorganisation in der Vormoderne" [Pre-Modern Clocks, Bells and Time Organization]. In *Städtische Repräsentation: St. Reinoldi und das Rathaus als Schauplätze des Dortmunder Mittelalters* [Urban Representation: St. Reinold's Church and the Town Hall as Scenes of Dortmund's Middle Ages], edited by Nils Büttner, Thomas Schilp, and Barbara Welzel, 59–77. Bielefeld: Verlag für Regionalgeschicht, 2005.

———. "Glocken und Uhren: Zeitmessung und Zeitordnungen in der Stadt" [Bells and Clocks: Time Measurement and Time Organization in the City]. In *Zeiten und Räme—Rhythmus und Region* [Time and Space—Rhythm and Region], edited by Dietmar Schiersner, 73–94. Constance: UVK Verlagsgesllschaft, 2013.

———. "Uhren und Uhrzeiten in den mittelalterlichen Städten" [Clocks and Clock Times in Cities of the Middle Ages]. In *Mittelalterliche astronomische Grossuhren* [Astronomical Great Clocks of the Middle Ages], edited by Manfred Schukowski, Uta Jahnke, and Wolfgang Fehlberg, 75–94. Leipzig, Akademische Verlagsanstalt, 2014.

Dohrn-van Rossum, Gerhard, and Rolf Westheider. "Die Einführung der öffentlichen Uhren und der Übergang zur modernen Stundentrechnung in den spätmittelalterlichen Städten Nieersachsens" [The Introduction of Public Clocks and the Transition to Modern Hour Reckoning in the Late Medieval Towns of Lower Saxony]. *Stadt im Wandel* [City in Transition], edited by Cord Meckseper, 317–31. Stuttgart-Bad Cannstatt: Edition Cantz, 1985.

Doppelmayr, Johann Gabriel. *Historische Nachricht von den Nürnbergischen Mathematicis und Künstlern* [Historical Accounts of Nuremberg's Mathematicians and Artists]. Nuremberg: Peter Conrad Monaths, 1730.

Drake, Francis. *Eboracum or, the History and Antiquities of the City of York*. London: William Bowyer, 1736.

Dreyer, I. L. E. "Prolegomena Editoris to Tychonis Brahe Dani Scripta Astronomica" [Editorial Prologue to Tycho Brahe's Danish Astronomical Writings]. In *Tychonis Brahe Dani Scripta Astronomica, Tomus I* [The Danish Astronomical Writings of Tycho Brahe, Volume 1], edited by I. L. E. Dreyer, V-LIX. Hauniae: Libraria Gyldendaliana, 1913.

Dunbar, R. I. M., A. H. Korstjens, and J. Lehmann. "Time as an Ecological Constraint." *Biological Reviews* 84, no. 3 (2009): 413–29.

Dunn, Elizabeth Cullen. "Standards without Infrastructure." In *Standards and Their Stories*, edited by Martha Lampland and Susan Leigh Star, 118–21. Ithaca: Cornell University Press, 2009.

Dunn, Marcia. "What Time Is It on the Moon? Europe Pushing for Lunar Time Zone." *Associated Press,* 28 February 2023. Retrieved 5 January 2026 from https://apnews.com/article/moon-time-zone-space-2b0124415c14755e08a58e1b5ed5362a.

Edgell, Beatrice, and W. Legge Symes. "The Wheatstone-Hipp Chronoscope: Its Adjustments, Accuracy, and Control." *British Journal of Psychology* 2, no. 1 (1906): 58–88.

Edmonson, Munro S. *The Book of the Year: Middle American Calendrical Systems.* Salt Lake City: University of Utah Press, 1988.

Edwards, Amelia. *A Thousand Miles Up the Nile.* Los Angeles: J. P. Tascher, 1983.

Einstein, Alfred. *Relativity*, translated by Robert W. Lawson. New York: Barnes and Noble, 2004.

Elliott, J. H. "A Europe of Composite Monarchies." *Past and Present* 137, no. 1 (1992): 48–71.

Epstein, Steven A. *Wage Labor and Guilds in Medieval Europe.* Chapel Hill, NC: University of North Carolina Press, 1991.

Esbester, Mike "Designing Time: The Design and Use of Nineteenth-Century Transport Timetables." *Journal of Design History* 22, no. 2 (2009): 91–113.

Essen, Louis. "Time Scales." *Metrologia* 48, no. 4 (1968): 161–65.

Essen, Ray. 2020. *Revolutions in Time.* (Independently published).

European Space Agency. "Telling Time on the Moon." 27 February 2023. Retrieved 27 June 2023 from https://www.esa.int/Applications/Navigation/Telling_time_on_the_Moon.

Evans, R. J. W. *Rudolf II and his World: A Study in Intellectual History 1576–1612.* Oxford: Clarendon, 1973.

———. Austrian Habsburgs: The Dynasty as a Political Institution. In *The Courts of Europe: Politics, Patronage and Royalty, 1400–1800*, edited by A. G. Dickens, 121–46. New York: McGraw-Hill, 1977.

———. *The Making of the Habsburg Monarchy 1550–1700.* Oxford: Clarendon, 1979.

Fabian, Johannes. *Time and the Other: How Anthropology Makes Its Object.* New York: Columbia University Press, 1983.

Febvre, Lucien. *The Problem of Unbelief in the Sixteenth Century, the Religion of Rabelais,* translated by Beatrice Gottlieb. Cambridge: Harvard University Press, 1982.

Felski, Rita. *Doing Time: Feminist Theory and Postmodern Culture.* New York: New York University Press, 2000.

Fernandez, M. P., and Fernandez, P. C. "Precision Timekeepers of Tokugawa Japan and the Evolution of the Japanese Domestic Clock." *Technology and Culture* 37, no. 2 (1996): 221–48.

Findlen, Paula. *Possessing Nature: Museums, Collecting, and Scientific Culture in Early Modern Italy.* Berkeley: University of California Press, 1994.

Finkleman, David, Steve Allen, John H. Seago, Rob Seaman, and P. Kenneth Seidelmann. "The Future of the Leap Second." *American Scientist* 99, no. 4 (2011): 312–19.

Fitzpatrick, Tony. "Social Policy and Time." *Time and Society* 13, nos. 2/3 (2004): 197–219.

Fosque, Hugh. "The International Radio Consultative Committee (CCIR): Its Role, Function and Influence on the Distribution of Time and Frequency Information." *Proceedings of the Precise Time and Time Interval (PTTI) Systems and Applications Meeting.* 8 (1976): 17–28.

Foucault, Michel. *Discipline and Punish: The Birth of the Prison*, translated by Alan Sheridan. London: Allen Lane, 1977.

———. *Power/Knowledge: Selected Interviews and Other Writings, 1972–1977, edited* and translated by Colin Gordon. New York: Vintage, 1980.

Foulkes, Nicholas. *Time Tamed.* London: Simon and Schuster, 2019.

Frake, Charles O. "Cognitive Maps of Time and Tide among Medieval Seafarers." *Man* 20, no. 2 (1985): 254–70.

Franklin, Allan. *The Trail of the Tiger.* New York: Allan Franklin, 1928.
Freeman, Elizabeth, *Time Binds: Queer Temporalities, Queer Histories* Durham, NC: Duke University Press, 2010.
Friedman, Batya, and Helen Nissenbaum. "Bias in Computer Systems." *ACM Transactions on Information Systems* 14, no. 3 (1996): 330–47.
Friedman, Thomas L. *The World Is Flat: A Brief History of the Twenty-First* Century. New York: Farrar, Straus and Giroux, 2005.
Frumer, Yulia. "Translating Time: Habits of Western-Style Timekeeping in Late Edo Japan," *Technology and Culture,* 55, no 4 (2014): 785–820.
———. *Making Time: Astronomical Time Measurement in Tokugawa Japan.* Chicago: University of Chicago Press, 2018.
Fučíková, Eliška. "The Belvedere in Prague as Tycho Brahe's Musaeum." *Tycho Brahe and Prague: Crossroads of European Science,* edited by J. R. Christianson, Alena Hadravová, Petr Hadrava, and Martin Šolc, 276–81. Frankfurt am Main: Verlag Harri Deutsch, 2002.
Fuller, A. W. "Universal Recilineal Dials." *The Mathematical Gazette* 41, no. 335 (1957): 9–24.
Gaab, Hans. "Die grosse Nürnbergische Uhr" [The Great Nurembergian Clock]. *Beiträge zur Astronomiegeschichte* [Comtributions to the History of Astronomy]. 8 (2006): 43–90.
Galison, Peter. *Einstein's Clocks, Poincaré's Maps.* New York: Norton, 2001.
Garcia-Diego, José A. *Juanelo Turriano: Charles V's Clockmaker, The Man and His Legend.* Nantuckett, MA: Science History Publications, 1986.
Garrioch, David. "Sounds of the City: The Soundscape of Early Modern European Towns." *Urban History* 30, no. 1 (2003): 5–25.
Gaulke, Karsten. "'The First European Observatory of the Sixteenth Century, as Founded by Langrave Wilhelm IV of Hesse-Kassel': A Serious Historiographic Category of a Misleading Marketing Device." In *European Collections of Scientific Instruments, 1550–1750,* edited by Giorgio Strano, Stephen Johnston, Mara Miniati, and Alison Morrison-Low, 87–99. Leiden: Brill, 2009.
Gayle, Damien. "Britain Battles U.S. to Keep the Leap Second at Conference which Could See 'Biggest Timekeeping' Change for Centuries." *Daily Mail,* 19 January 2012.
Gill, Patrick. "When Should We Change the Definition of the Second." *Philosophical Transactions of the Royal Society A: Mathematical, Physical and Engineering Sciences* 369, no. 1953 (2011): 4,109–30.
Gimpel, Jean. *The Medieval Machine: The Industrial Revolution of the Middle Ages.* New York: Penguin, 1976.
Gingerich, Owen. *The Eye of Heaven: Ptolemy, Copernicus, Kepler.* New York: American Institute of Physics, 1993.
Glennie, Paul, and Nigel Thrift. *Shaping the Day: A History of Timekeeping in England and Wales 1300–1800.* Oxford: Oxford University Press, 2009.
Glissant, Éduoard. *Caribbean Discourse,* translated by J. Michael Dash. Charlottesville, VA: University of Virginia Press, 1989.
———. 1997. *Poetics of Relation,* translated by Betsy Wing. Ann Arbor: University of Michigan Press.
Globe. "Curtis v. :" 23 July 1858: 4.
Godun, Rachel M. "Atomic Clocks Compared with Astounding Accuracy." *Nature* 591 (2021): 534–35.
Goethe, Johann Wolfgang von. *Italian Journey,* translated by W. H. Auden and Elizabeth Mayer. New York: Penguin, 1970.

Goldman, David. "This Year's Y2K: 'Leap Second' Threatens to Break the Internet." *CNNMoney*, 13 January 2015.

Golinski, Jan. *Making Natural Knowledge: Constructivism and the History of Science.* Cambridge: Cambridge University Press, 1998.

Goodman, David C. *Power and Penury: Government, Technology and Science in Philip II's Spain.* Cambridge: Cambridge University Press, 1988.

Gordon, Geoff. "Railway Clocks." In *International Law's Objects*, edited by Jessie Hohmann and Daniel Joyce, 387–98. Oxford: Oxford University Press, 2018.

———. "Imperial Standard Time." *The European Journal of International Law.* 29, no. 4 (2019): 1,197–222.

Grabham, Emily. *Brewing Legal Times: Things, Form and the Enactment of Law.* Toronto: University of Toronto Press, 2016.

Green, Nicola. "On the Move." *The Information Society* 18, no. 4 (2002): 281–92.

Greenhouse, Carol. *A Moment's Notice.* Ithaca, NY: Cornell University Press, 1996.

Griffin, Andrew. "Computer Chaos feared over 2015's leap second." *USA Today*, 8 January 2015. Retrieved 23 September 2021 from http://www.usatoday.com/story/tech/2015/01/08/computer-chaos-feares/21433363/.

Griffiths, Jay. *A Sideways Look at Time.* New York: Jeremy P. Tarcher/Putnam, 1999.

Groebner, Valentin. *Liquid Assets, Dangerous Gifts: Presents and Politics at the End of the Middle Ages*, translated by Pamela E. Selwyn. Philadelphia: University of Pennsylvania Press, 2000.

Grössing, Helmuth. "*Regiomontanus und Italien: Zum Problem der Wissenschaftsauffassung des Humanismus*" [Regiomontanus and Italy: On the Problem of the Humanist Conception of Science]. In *Regiomontanus-Studien* [Regiomontanus Studies], edited by Hamann Günther, 223–41. Vienna: Verlag der Österreichischen Akademie der Wissenschaften, 1980.

Guéna, J., M. Abgrall, A. Clairon, and S. Bize. "Contributing to TAI with a Secondary Representation of the SI Second." *Metrologia* 51, no. 1 (2014): 108–20.

Guinot, Bernard. "Review of Time Scales." *6th Annual PTTI Meeting* (1974): 17–26.

———. "Some Properties of Algorithms for Atomic Time Scales." *Metrologia* 24, no. 4 (1987): 195–98.

Guinot, B., and P. K. Seidelmann. "Time Scales: Their History, Definition and Interpretation." *Astronomy and Astrophysics* 194, nos. 1–2 (1988): 304–8.

Gupta, Akhil. "The Reincarnation of Souls and the Rebirth of Commodities." *Cultural Critique* 22 (1992): 187–211.

Gürbüz, Şule. "Turkish Time of Yesteryear." In *Visible Faces of the Time: Timepieces*, edited by Sennur Sentürk and Selahattin Özpalabiyiklar, 57–66. Istanbul: YKY, 2009.

———. "Ezani Time and the Islamic Perception of Time." In *Topkapi Palace Clock Collection: Exceptional Timepieces at the Seraglio*, translated by Nazim Dikbas, 10–17. Istanbul: Promat Basim Yayin San, 2013.

Gustafsson, Harald. "The Conglomerate State: A Perspective on State Formation in Early Modern Europe." *Scandinavian Journal of History* 23, nos. 3–4 (1998): 189–213.

Hacking, Ian. "The Self-Vindication of the Laboratory Sciences". In *Science and Practice and Culture*, edited by Andrew Pickering, 29–64. Chicago: University of Chicago Press, 1992.

Hafele, J. C., and Richard Keating. "Around-the-World Atomic Clocks: Observed Relativistic Time Gains." *Science* 177, no. 4044 (1972): 168–70.

Hall, Richard. "The Earliest Record of a Clock at York Minster?" *Seventy Ninth Annual Report*, 69–73. York: Friends of York Minster, 2008.

Hamann, Günther. "Johannes Regiomontanus 1436–1476: Die Schauplätze seines Lebens und Wirkens" [Johannes Regiomontanus 1436–1476: The Setting of his Life and Work].

In *Regiomontanus-Studien* [Regiomontanus Studies], edited by Hamann Günther, 11–46. Vienna: Verlag der Österreichischen Akademie der Wissenschaften, 1980.

Hamburger, Jeffrey. "Introduction: Histories of Female Monasticism." In *Crown and Veil: Female Monasticism from the Fifth to the Fifteenth Centuries*, edited by Jeffrey Hamburger and Susan Marti, translated by Dietlinde Hamburger, 1–11. New York: Columbia University Press, 2008.

Hampson, Robert Thomas. *Medii Aevi Kalendarium* [The Medieval Calendar]. London: Henry Kent Causton and Co., 1841.

Harley, J. Brian. "Silences and Secrecy: The Hidden Agenda of Cartography in Early Modern Europe". *Imago Mundi* 40, no. 1 (1988): 57–76.

———. "Rereading the Maps of the Columbian Encounter." *Annals of the Association of American Geographers* 82, no. 3 (1992): 522–36.

Harris, Wilson. *Tradition, the Writer and Society: Critical Essays*. London: New Beacon, 1967.

Harrison, Brian. *Drink and the Victorians: The Temperance Question in England, 1815–1872*. Pittsburgh: University of Pittsburgh Press, 1971.

Harvey, David. *The Condition of Postmodernity*. Oxford: Blackwell, 1989.

———. "Between Space and Time: Reflections on the Geographical Imagination." *Annals of the Association of American Geographers* 80, no. 3 (1990): 418–34.

———. "From Space to Place and Back Again: Reflections on the Condition of Postmodernity." In *Mapping the Futures: Local Cultures, Global Change*, edited by Jon Bird, Barry Curtis, Tim Putnam, George Robertson, and Lisa Tickner, 3–29. London: Routledge, 1993.

Hassan, Robert. *The Chronoscopic Society*. New York: Peter Lang, 2003.

———. Timescapes of the Network Society. *Fast Capitalism* 1, no. 1 (2005): 23–32.

———. *The Information Society*. Cambridge: Polity Press, 2008.

———. "Analogue Time, Analogue People and the Digital Eclipsing of Modern Political Time." In *Time, Temporality and Global Politics*, edited by Andrew Hom, Christopher McIntosh, Alasdair McKay, and Liam Stockdale, 87–101. Bristol: E-International Relations Publishing, 2016.

———. "Time and Sovereignty in the Neoliberal Hegemony." In *Time, Globalization and Human Experience: Interdisciplinary Explorations*, edited by Paul Huebener, Susie O'Brien, Tony Porter, Liam Stockdale, and Yanqiu Rachel Zhou, 26–46. London: Routledge, 2017.

———. *Uncontained: Digital Disconnection and the Experience of Time*. Grattan Street Press, 2019.

Hayton, Darin. "Expertise ex Stellis: Comets, Horoscopes, and Politics in Renaissance Hungary." *Osiris* 25 no. 1 (2010): 27–46.

Heal, Felicity. *The Power of Gifts: Gift Exchange in Early Modern England*. Oxford: Oxford University Press, 2014.

Heidegger, Martin. *The Question Concerning Technology and Other Essays*, translated by W. Lovitt. New York: Harper and Row, 1977.

Heilbron, J. L. *The Sun in the Church: Cathedrals as Solar Observatories*. Cambridge, MA: Harvard University Press, 1999.

Henderson, John Tasker. "The Foundation of Time and Frequency in Various Countries."
Proceedings of the IEEE 60, no. 5 (1972): 487–93.

Henzi, S. P., R. W. Byrne, and A. Whiten. "Patterns of Movement by Baboons in the Drakensberg Mountains: Primary Responses to the Environment." *International Journal of Primatology* 13, no. 6 (1992): 601–29.

Hervey, Christopher. *Letters from Portugal, Spain, Italy and Germany, in the Years 1759, 1760, and 1761, volume 2*. London: J. Davis, 1785.

Hope, Wayne. "Conflicting Temporalities." *Time and Society* 18, no. 3 (2009): 62–85.
———. *Time, Communication, and Global Capitalism.* New York: Palgrave Macmillan, 2016.
———. "Time, Worker Exploitation, and Global Capitalism." In *Time, Globalization and Human Experience: Interdisciplinary Explorations*, edited by Paul Huebener, Susie O'Brien, Tony Porter, Liam Stockdale, and Yanqiu Rachel Zhou, 47–67. London: Routledge, 2017.
Hörnung, Karl H., Daniela Ahrens, and Anette Gerhard, Anette. "Do Technologies Have Time?" *Time and Society* 8, no. 2–3 (1999): 293–308.
Hoskins, Janet. *The Play of Time.* Berkeley: University of California Press, 2007.
Howard, Deborah, and Laura Moretti. *Sound and Space in Renaissance Venice: Architecture, Music, Acoustics.* New Haven: Yale University Press, 2009.
Howse, Derek. *Greenwich Time and the Discovery of Longitude.* Oxford: Oxford University Press, 1980.
Hudson, George E. "Some Characteristics of Commonly Used Time Scales." *Proceedings of the IEEE* 55, no. 6 (1967): 815–21.
Hudson, George E., and William Atkinson. "The Redefinition of the Second and the Velocity of Light." *Physics Today* 16, no. 5 (1963): 30–36.
Huebener, Paul, Susie O'Brien, Tony Porter, Liam Stockdale, and Yaniu Rachel Zhou. "Exploring the Intersection of Time and Globalization." *Globalizations* 13, no. 3 (2016): 243–55.
———. "Introduction." In *Time, Globalization and Human Experience: Interdisciplinary Explorations*, edited by Paul Huebener, Susie O'Brien, Tony Porter, Liam Stockdale, and Yanqiu Rachel Zhou, 1–25. London: Routledge, 2017.
Hughes, Paul. "Implicit Carolingian Tidal Data. "*Early Science and Medicine* 8 no. 1 (2003): 1–24.
Humphrey, Chris. "Time and Urban Culture in Late Medieval England." In *Time in the Medieval World*, edited by Chris Humphrey and W. M. Ormrod, 105–18. York: York Medieval Press, 2001.
Hurlstone, E. T., and J. P. Norman. *The Exchequer Reports, Volume II.* Philadelphia: T. and J.W. Johnson and Company, 1858.
Hutchins, Edwin. *Cognition in the Wild.* Cambridge, MA: MIT Press, 1995.
———. "Cognitive Artifacts." In *MIT Encyclopedia of the Cognitive* Sciences, edited by Robert A.Wilson and Frank C. Keil, 126–27. Cambridge, MA: MIT Press, 1999.
———. "The role of Cultural Practices in the Emergence of Modern Human Intelligence." *Philosophical Transactions of the Royal Society, B*, 363, no. 1499 (2011–2019).
———. "The Cultural Ecosystem of Human Cognition." *Philosophical Psychology* 27, no. 1 (2014): 34–49.
IAU. "The 1952 General Assembly of the IAU." *Transactions of the International Astronomical Union* 8 (1954).
———. "Commission de l'Heure." *Transactions of the International Astronomical Union—Reports* 11A (1962): 361–368.
———. "Resolutions Adopted by the General Assembly." *Transactions of the International Astronomical Union* 13, no. 2 (1967): 37–43.
———. "Time." *Transactions of the International Astronomical Union* 14, no. 1 (1970): 343–55.
IEEE Standards Association. *IEEE Standard for a Precision Clock Synchronization Protocol for Networked Measurement and Control Systems Amendment 2: Master-Slave Optional Alternative Terminology*, 2022.

IEEE Standards Association. *IEEE Standard for Use of Inclusive Language in Technical Terminology and Communications*, 2025.

Ihde, Don. *Instrumental Realism: The Interface between Philosophy of Science and Philosophy of Technology*. Bloomington, IN: Indiana University Press, 1991.

———. *Postphenomenology: Essays in the Postmodern Context.* Evanston, IL: Northwestern University Press, 1993.

International Meridian Conference. *International Conference Held at Washington for the Purpose of Fixing a Prime Meridian and a Universal Day: Protocols of the Proceedings.* Washington, DC: Gibson Bros, 1884.

Isidore of Seville. *Etymologarium. Patrologia Latina, volume 82*, edited by J.–P. Migne, columns 73–728. Turnhout: Typographi Brepols Editores Pontifici, 1844–1864.

ISO. *Economic Benefits of Standards, volume 1*. Geneva: ISO Central Secretariat, 2011.

———. *Economic Benefits of Standards, volume 2: International Case Studies.* Geneva: ISO Central Secretariat, 2012.

———. *Economic Benefits of Standards*. Geneva: ISO Central Secretariat, 2014.

ITU. *Radio Regulations: Additional Radio Regulations, Resolutions and Recommendations.* Geneva: General Secretariat of the International Telecommunication Union, 1968.

———. "XII Plenary Assembly of the CCIR." *Telecommunication Journal* 36, no. 5 (1970): 207–12.

———. "Coordinated Universal Time (UTC) to retain 'Leap Second.'" 19 November 2015. Retrieved 19 November 2015 from http://www.itu.int/net/pressoffice/press_releases/2015/53.aspx#.W32l6H4h2i4.

Jäger, Edmund. "Einleitung in die historische chronologie" [Introduction to Historical Chronology]. *Korrespondenz-Blatt für die Gelehrten- und Realschulen Württembergs* [Bulletin for the Academic and Secondary Schools of Württemberg]. July–August, 295–329. Tübingen: Franz Fues, 1881.

James, C. L. R. *Wilson Harris: A Philosophical Approach.* Port of Spain: University of the West Indies, 1965.

James, William. *Principles of Psychology*. New York: Dover, 1950.

Jameson, Fredric. *Postmodernism or, the Cultural Logic of Late Capitalism.* Duke: Duke University Press, 1991.

Jami, Catherine. "Western Devices for Measuring Time and Space: Clocks and Euclidian Geometry in Late Ming and Ch'ing China." In *Time and Space in Chinese Culture*, edited by Chun-Chieh Huang and Erik Zürcher, 169–200. Leiden: Brill, 1995.

Johns, Fleur. "The Temporal Rivalries of Human Rights." *Indiana Journal of Global Legal Studies* 23, no. 1 (2016): 39–60.

Johnson, Barbara. *Persons and Things.* Cambridge, MA: Harvard University Press, 2008.

Johnson, Thomas, Stephen Malys, and Barbara Wiley. "The GPS SVN59 Satellite Anomaly of 17 June 2012." In *Requirements for UTC and Civil Timekeeping on Earth*, edited by J. H. Seago, R. L. Seaman, P. K. Seidelmann, and S. L. Allen. American Astronautical Society, Science and Technology Series, vol. 115, 101–4. San Diego: Univelt, 2013.

Johnson, Willis Fletcher, and Ray B. Smith. *Political and Governmental History of the State of New York.* Syracuse: Syracuse Press, 1922.

Jones, Basil, ed. *The American Ruling Cases as Determined by the Courts, volume I.* Chicago: National Law Book Company, 1912.

Jones, Charles. 1943. "Development of the Latin Ecclesiastical Calendar." In *Bedae Opera de Temporibus* [Bede's Works about Time], edited by Charles Jones, 3–121. Cambridge, MA: The Mediaeval Academy of America, 1943.

Kanigel, Robert. *One Best Way: Frederic Winslow Taylor and the Enigma of Efficiency.* New York: Viking, 1997.

Kaplan, George H. "Time Scales in Astronomical and Navigational Almanacs." In *Decoupling Civil Timekeeping from Earth Rotation*, edited by John H. Seago, Robert L. Seaman, and Steven L. Allen. American Astronautical Society, Science and Technology Series, vol. 113, 201–14. San Diego: Univelt, 2011.

Kaufmann, Thomas DaCosta. *Variations on the Imperial Theme in the Age of Maximilian II and Rudolf II*. New York: Garland, 1978.

———. "Remarks on the Collections of Rudolf II: The *Kunstkammer* as a form of *Representatio*." *Art Journal* 38, no. 1 (1978): 22–28.

Keightley, Emily. "From Immediacy to Intermediacy." *Time and Society* 22, no. 1 (2013): 55–75.

Keyssler, Johann Georg. *Travels through Germany, Bohemia, Hungary, Switzerland, Italy, and Lorrain*, v. 4. London: A. Lind, 1756–1757.

King, David A. "The Astrolabe Presented by Regiomontanus to Cardinal Bessarion in 1462." *Nuncius* 9, no. 1 (1994): 165–205.

———. *Astrolabes and Angels, Epigrams and Enigmas*. Stuttgart: Franz Steiner Verlag, 2007.

Kircher, Athanasius. *Phonurgia Nova sive Coniugium Mechanico-physicum artis et naturae* [The New Method of Sound Production, or the Mechanical-Physical Marriage of Art and Nature]. Campidona: Rudolph Dreherr, 1673.

Knight, Charles Brunton. *A History of the City of York*. York: Herald Printing Works, 1944.

Koenigsberger, H. G. *The Habsburgs and Europe, 1516–1660*. Ithaca, NY: Cornell University Press, 1971.

———. *Early Modern Europe 1500–1789*. Longman: London, 1987.

Kovalevsky, Jean, and Terrence J. Quinn. "The International System of Units (SI)." *Comptes Rendus Physique* [Physics Reports] 5 (2004): 799–811.

Krislov, Samuel. *How Nations Choose Product Standards and Standards Change Nations*. Pittsburgh: University of Pittsburgh Press, 1997.

Lalande, Joseph Jérôme le Français de. *Voyage d'un François en Italie, fait dans les années 1765 & 1766*. Paris: Desaint, 1769.

Landes, David. *Revolution in Time: Clocks and the Making of the Modern World*. Cambridge, MA: Belknap, 1983.

Lanfranc. *The Monastic Constitutions*, translated by David Knowles. London: Thomas Nelson. 1951.

Lanham, Richard. *The Electronic Word: Democracy, Technology and the Arts*. Chicago: University of Chicago Press, 1994.

Latour, Bruno. *Science in Action: How to Follow Scientists and Engineers Through Society*. Cambridge: Harvard University Press, 1987.

———. "Technology as Society Made Durable." In *A Sociology of Monsters: Power, Technology and Domination*, edited by John Law, 103–31. London: Routledge, 1991.

———. *We Have Never Been Modern*, translated by Catherine Porter. Cambridge, MA: Harvard University Press, 1993.

———. Actor-Network Theory: A Few Clarifications. *Soziale Welt* 47, no. 4 (1996): 369–81.

———. *Aramis or the Love of Technology*, translated by Catherine Porter. Cambridge: Harvard University Press, 1996.

———. *Reassembling the Social: An Introduction to Actor-Network Theory*. Oxford: Oxford University Press, 2005.

Latour, Bruno, and Steve Woolgar. *Laboratory life: The Construction of Scientific Facts*, 2nd edn. Princeton: Princeton University Press, 1986.

Lazar, Nomi Claire. *Out of Joint: Power, Crisis, and the Rhetoric of Time*. New Haven: Yale University Press, 2019.

Leakey, Louis, P., V. Tobias, and J. R. Napier. "A New Species of the Genus *Homo* from Olduvai Gorge." *Nature* 202 (1964): 7–9.

Le Goff, Jacques. *Time, Work, and Culture in the Middle Ages*, translated by Arthur Goldhammer. Chicago: University of Chicago Press, 1980.

———. *Medieval Civilization, 400–1500*, translated by Julia Barrow. Oxford: Blackwell, 1988.

Le Matin. "Quelle heure est-il? ou l'impossibilité d'une réponse exacte" [What hour is it? Or the Impossibility of an Exact Answer]. 13 August 1939: 2.

Lenzen, Victor F., and Robert P. Multhauf. "Gravity Pendulums in the 19th Century." *United States National Museum Bulletin* 240 (1966): 303–47.

Leroi-Gourhan, André. *Gesture and Speech*. Cambridge, MA: MIT Press, 1993.

Levine, Judah. "The Future of Time and Frequency Dissemination." *Proceedings of the 1993 PTTI Meeting*, 573–78.

———. "Introduction to Time and Frequency Metrology." *Review of Scientific Instruments* 70, no. 6 (1999): 2,567–96.

———. "Time and Frequency Distribution Using Satellites." *Reports on Progress in Physics* 65, no. 8 (2002): 1,119–64.

———. "The Statistical Modeling of Atomic Clocks and the Design of Time Scales." *Review of Scientific Instruments* 83, no. 021101 (2012): 1–28.

———. "Time Transfer Using a Satellite Navigation System." *McGraw Hill Yearbook of Science and Technology* (2013), 363–68.

———. "The History of Time and Frequency from Antiquity to the Present Day." *The European Physical Journal H*. 41, no. 1 (2016): 1–67.

Levine, Judah, Patrizia Tavella, and Martin Milton. "Towards a Consensus on a Continuous Coordinated Universal Time." *Metrologia* 60, no. 1 (2023): 1–9.

Levine, Robert. *A Geography of Time: The Temporal Misadventures of a Social Psychologist*. New York: Basic Books, 1997.

Lewandowski, W. "Global Navigation Satellite Systems and their System Times." *ITU News*, January (2014). https://itunews.itu.int/en/NotePrint.aspx?Note=4272.

Lewandowski, W., and E. F. Arias. "GNSS Times and UTC." *Metrologia* 48, no. 4 (2011): S219–24.

Lewandowski, W., D. Matsakis, G. Panfilo, and P. Tavella. "The Evaluation of Uncertainties in [UTC – UTC(k)]." *Metrologia* 43 (2006): 278–86.

l'Hermite, Jehan. *Le Passetemps, volume 2* [The Hobby], edited by Émile Ouverleaux and Jules-Jean Petit. Geneva: Slatkine, 1896.

Lindgren, Uta. 2010. "Die Bedeutung von Regiomontans Ephemeriden für die Entdecung Amerikas" [The Importance of Regiomontanus' Ephemerides for the Discovery of America]. In *Astronomie in Nürnberg* [Astronomy in Nuremberg], edited by Gudrun Wolfschmidt, 145–56. Hamburg: Tredition Science, 2010.

Lloyd, H. Alan. *Some Outstanding Clocks Over Seven Hundred Years, 1250–1950*. London: L. Hill, 1958.

Loconto, Allison, and Lawrence Busch. "Standards, Techno-economic Networks, and Playing Fields: Performing the Global Market Economy." *Review of International Political Economy* 17, no. 3 (2010): 507–36.

de Longh, Jane. *Margaret of Austria: Regent of the Netherlands*, translated by M. D. Herter Norton. New York: W. W. Norton, 1953.

Los Angeles Times. "Wait A Second—Time Will Stand Still Today." 30 June 1972, part II: 5.

Luhrmann, Tanya M. "The Magic of Secrecy." *Ethos* 17, no. 2 (1989): 131–65.

Lynch, Michael. Representation Is Overrated: Some Critical Remarks about the Use of the Concept of Representation in Science Studies. *Configurations* 2, no. 1 (1994), 137–49.

Machiavelli, Nicolò. *The Prince.* Translated by Peter Bondanella and Mark Musa. Oxford: Oxford University Press, 1984.

Malys, Stephen. 2011. "Proposal for the Redefinition of UTC: Influence on NGA Earth Orientation Predictions and GPS Operations." In *Decoupling Civil Timekeeping from Earth Rotation*, edited by John H. Seago, Robert L. Seaman, and Steven L. Allen. American Astronautical Society, Science and Technology Series, vol. 113, 265–70. San Diego: Univelt, 2011.

Manchester Evening News. "Publicans and Closing Time." 1 January 1878: 4.

Maniewicz, Mario. "The Future of Timekeeping." *ITU News Magazine* 2 (2023): 6–8.

Marpergers, Paul Jacob. *Horologiographia* [Horology]. Munich: Georg D. W. Callwey, 1978.

Marshall, Peter. *The Mercurial Emperor: The Magic Circle of Rudolf II in Renaissance Prague.* London: Pimlico, 2007.

Masselos, Jim. "Bombay Time." In *Intersections,* edited by Meera Kosambi, 161–83. New Delhi: Orieng Longman, 2000.

Massey, Doreen. *Space, Place and Gender.* Cambridge: Polity Press, 1994.

Mathematisch-Physikaler Salon. *Uhren-Globen: wissenschaftliche Instrumente* [Clock-Globes: Scientific Instruments]. Dresden: Karl M. Lipp, 1993.

Matsakis, Demetrios. "A Generalizable Formalism for Computing Uncertainties in UTC." *Metrologia* 57, no. 6 (2020). DOI 10.1088/1681-7575/aba228.

Matsakis Demetrios, Judah Levine, and M. A. Lombardi. "Metrological and Legal Traceability of Time Signals." *Proceedings of the 49th Annual Precise Time and Time Interval Systems and Applications Meeting (PTTI)* (2018): 59–71.

Maurice, Klaus. "Jost Bürgi, or On Innovation". In *The Clockwork Universe: German Clocks and Automata 1550–1650,* edited by Klaus Maurice and Otto Mayr, 87–102. Washington, DC: Smithsonian, 1980.

Maury, Mytton. "Martin Behaim's Globe, and His Influence upon Geographical Science." *Journal of the American Geographical Society of New York* 4 (1873): 432–52.

Mawani, Renisa. "Law as Temporality: Colonial Politics and Indian Settlers." *U.C. Law Review* 4, no. 1 (2014): 65–95.

Mbembe, Achille. *On the Postcolony.* Berkeley: University of California Press, 2001.

McCarthy, Dennis. "Precision Time and the Rotation of the Earth." *Transits of Venus: New Views of the Solar System and Galaxy.* Proceedings IAU Colloquium 196 (2004): 180–96.

———. "Note on Coordinated Universal Time (CCTF/09-32)." *Consultative Committee for Time and Frequency Report of the 18th Meeting (4–5 June 2009) to the International Committee for Weights and Measures.* Sévre: BIPM.

———. "Evolution of Timescales from Astronomy to Physical Metrology." *Metrologia* 48, no. 4 (2011): S132–44.

———. "Considering Time-Scale Requirements for the Future." In *Requirements for UTC and Civil Timekeeping on Earth,* edited by J. H. Seago, R. L. Seaman, P. K. Seidelmann, and S. L. Allen. S. L. American Astronautical Society, Science and Technology Series, vol. 115, 301–18. San Diego: Univelt, 2013.

McCarthy, Dennis, Christine Hackman, and Robert Nelson. "The Physical Basis of the Leap Second." *The Astronomical Journal* 136, no. 5 (2008): 1,906–8.

McCarthy, D., and W. J. Klepczynski. "GPS and Leap Seconds: Time to Change?" *GPS World.* November (1999): 50–54.

McCarthy, Dennis D., and P. Kenneth Seidelmann. *Time—From Earth Rotation to Atomic Physics.* Weinheim: Springer, 2009.

McCrossen, Alexis. "Conventions of Simultaneity." *Journal of Urban History* 33, no. 2 (2007): 217–53.

———. "The 'Very Delicate Construction' of Pocket Watches and Time Consciousness in the Nineteenth-Century United States." *Winterthur Portfolio* 44, no. 1 (2010): 1–30.

———. *Marking Modern Times: A History of Clocks, Watches and Other Timekeepers in American Life*. Chicago: University of Chicago Press, 2013.

———. 2020. "Broken Watches." In *Material Histories of Time*, edited by Gianenrico Bernasconi and Susanne Thürigen, 201–10. Berlin: Walter de Gruyter, 2020.

McGuigan, Dorothy Gies. *The Habsburgs*. Garden City, NY: Doubleday, 1966.

McGuire, James K. *The Democratic Party of the State of New York*, vol. 1. New York: United States History Company, 1905.

Mehta, Pherozeshah. "Standard Time vs. Bombay Time." In *Some Unpublished & Later Speeches & Writings of Pherozeshah Mehta*, edited by J. R. B. Jeejeebhoy, 171–81. Bombay: The Commercial Press, 1918.

Melancthon, Philipp. Preface to *De Judiciis Nativitatum* [The Judgements of Birth] by Johannes Schöner. Nuremberg: Montanus and Neuber, 1545.

Menzhausen, Joachim. "Elector Augustus' *Kunstkammer*: An Analysis of the Inventory of 1587." In *The Origins of Museums: The Cabinet of Curiosities in Sixteenth- and Seventeenth Century Europe*, edited by Oliver Impey and Arthur MacGregor, 91–100. Thirsk, North Yorkshire: House of Stratus, 2001.

Mignet, M. *Charles-Quint: son abdication, son séjour et sa mort au monastère de Yuste* [Charles V: His Abdication, His Stay and His Death at the Monastery of Yuste]. Paris: Paulin, L'heureux, et cie, 1854.

Miller, Daniel. *Material Culture and Consumption.* Oxford: Basil Blackwell, 1987.

Miller, E. "Medieval York." In *A History of Yorkshire: The City of York*, edited by P. M. Tillott, 25–116. London: Oxford University Press, 1961.

Miller, Edward, and Hatcher, John. *Medieval England: Towns, Commerce, and Crafts, 1086–1348*. London: Longman, 1995.

Mills, David. *Computer Network Time Synchronization: The Network Time Protocol on Earth and Space*, 2nd edn. Boca Raton: CRC Press, 2011.

Mody, N. H. N. *Japanese Clocks*. Rutland: Charles E. Tuttle, 1967.

Moran, Bruce T. "Machines and the Valuation of Precision in the 16th Century." *Sudhoffs Archiv* 61, no. 3 (1977): 209–28.

———. "Wilhelm IV of Hesse-Kassel: Informal Communication and the Aristocratic Context of Discovery." In *Scientific Discovery: Case Studies*, edited by Thomas Nickles, 67–98. Dordrecht, Holland: D. Reidel, 1980.

———. "German Prince-Practitioners: Aspects in the Development of Courtly Science, Technology, and Procedures in the Renaissance." *Technology and Culture* 22, no. 2 (1981): 253–74.

Moraw, Peter. "The Court of the German King and of the Emperor and the End of the Middle Ages." In *Princes, Patronage, and the Nobility: The Court at the Beginning of the Modern Age c. 1450–1650*, edited by Ronald G. Asch and Adolf M. Birke, 103–38. Oxford: Oxford University Press, 1991.

Morning Advertiser. "Dismissal of Two Summonses Against a Licensed Victualler." 2 February 1871: 6.

Morrison, J. E. *The Astrolabe*. Rehoboth Beach, DE: Janus, 2007.

Moyer, G. "Aloisius Lilius and the *Compendium Novae Rationis Restituendi Kalendarium.*" In *Gregorian Reform of the Calendar*, edited by G. V. Coyne, M. A. Hoskin, and O. Pederson, 171–88. Vatican City: Specola Vatican, 1983.

Mukerji, Chandra. "Visual Language in Science and the Exercise of Power: The Case of Cartography in Early Modern Europe." *Studies in Visual Communication* 10, no. 3 (1984): 30–45.

Mummenhoff, Ernst. "Lutz Steinlingers Baumeisterbuch vom Jahre 1452" [Lutz Steinlinger's Master Builder's Book from the Year 1452]. *Mitteilungen des Vereins für Geschichte der Stadt Nürnberg* [Reports of the Association for the History of the City of Nuremberg] 2 (1880): 15–77.

Murphy, Neil. "Receiving Royals in Later Medieval York: Civic Ceremony and the Municipal Elite, 1478–1503." *Northern History* 43, no. 2 (2006): 241–54.

Mushkat, Jerome. *Fernando Wood: A Political Biography*. Kent, Ohio: Kent State University Press, 1990.

Nanni, Giordano. *The Colonisation of Time: Ritual, Routine and Resistance in the British Empire*. Manchester: University of Manchester Press, 2013.

NAVSTAR *GPS User Equipment Introduction*. Arlington, VA: US Coast Guard, 1996.

NBS. "Atomic Clocks: First Class International Travelers." *NBS Technical New Bulletin* 56 (1972): 233.

Needham, Joseph. *Science and Civilisation in China, Volume 3: Mathematics and the Sciences of the Heavens and the Earth*. Cambridge: Cambridge University Press, 1959.

———. *Science and Civilization in China, volume 4: Physics and Physical Technology, part II: Mechanical Engineering*. Cambridge: Cambridge University Press, 1965.

Needham, Joseph, Wang Ling, and Derek de Solla Price. *Heavenly Clockwork: The Great Astronomical Clocks of Medieval China*. Cambridge: Cambridge University Press, 1960.

Nelson, Glenn K., Michael A. Lombardi, and Dean T. Okayama. *NIST Time and Frequency Radio Stations: WWV, WWVH, and WWVB*. Boulder, CO: National Institute of Standards and Technology, 2005.

Nelson, Robert A. "Relativistic Time Transfer in the Vicinity of the Earth and in the Solar System." *Metrologia* 48, no. 4 (2011): S171–80.

Nelson, R. A., D. D. McCarthy, S. Malys, J. Levine, B. Guinot, H. F. Fliegel, R. L. Beard, and T. R. Bartholmew. "The Leap Second: Its History and Possible Future." *Metrologia* 38, no. 6 (2001): 509–29.

Newcomb, Simon. *Astronomical Papers Prepared for the Use of the American Ephemeris and Nautical Almanac, volume VI, part I: Tables of the Sun*. Washington, DC: US Government Printing Office, 1895.

Newton, Isaac. *Sir Isaac Newton's Mathematical Principles of Natural Philosophy and his System of the World*, edited by Florian Cajori and translated by Andrew Motte. Berkeley: University of California Press, 1934.

New York Daily Tribune. "The Democracy at Syracuse." 19 September 1859: 4.

New York Herald. "The Collision on the Camden and Amboy Road." 11 August 1853, morning edition: 1.

———. "Municipal Affairs." 14 February 1856: 1.

———. "What Was Done in the Convention." 15 September 1859: 3.

———. "Details of the Proceedings of the Convention." 15 September 1859: 3.

New York Sun. "Only 8 Watches instead of 55." 13 May 1883: 3.

New York Times. "Message from the Mayor." 14 February 1856: 8.

———. "A Standard of Time." 16 February 1856: 4.

———. "Astronomical Time for the Port." 3 September 1856: 4.

———. "Proceedings of the Convention." 15 September 1859: 1.

———. "The Row at Syracuse: More Affidavits—Messrs. Carpenter and Ferris on the Stand." 17 October 1859: 5.

———. 1972. "Two Seconds to Make 1972 the Longest Year." 28 December 1972: 29.

North, J. D. "The Western Calendar— 'Intoleribilis, Horribilis, et Derisibilis': Four Centuries of Discontent." In *Gregorian Reform of the Calendar*, edited by G. V. Coyne, M. A. Hoskin, and O. Pederson, 75–113. Vatican City: Specola Vaticana, 1983.

———. *The Ambassadors' Secret: Holbein and the World of the Renaissance*. New York: Hambledon and London, 2004.

———. *God's Clockmaker Richard of Wallingford and the Invention of Time*. London: Hambledon and London, 2005.

North Wales Chronicle. "Wit and Humor." Saturday, November 11, Issue 1444 (1854): 7. In *British Library Newspapers, Part I: 1800–1900*. Retrieved 15 December 2025 from *British Library Newspapers*. https://link-gale-com.ezproxy.gc.cuny.edu/apps/doc/BB3200885061/BNCN?u=cuny_gradctr&sid=bookmark-BNCN&pg=7&xid=8b896814.

Nowotny, Helga. *Time: The Modern and Postmodern Experience*, translated by Neville Plaice. Cambridge: Polity Press, 1994.

O'Connell, Joseph. "Metrology: The Creation of Universality by the Circulation of Particulars." *Social Studies of Science* 23, no. 1 (1993): 129–73.

Ogle, Vanessa. *The Global Transformation of Time, 1870–1950*. Harvard: Harvard University Press, 2015.

OPM Group. *Leap Seconds Dialogue: Final Report*. London: OPM, 2014.

Orme, Nicholas. "Church and Chapel in Medieval England." *Transactions of the Royal Historical Society* 6 (1996): 75–102.

Ortiz, Fernando. *Cuban Counterpoint: Tobacco and Sugar*, translated by Harriet de Onís. Durham, NC: Duke University Press, 1995.

Pacey, Arnold. *The Culture of Technology*. Cambridge, MA: MIT Press, 1983.

Page, William, ed. *The Inventories of Church Goods for the Counties of York, Durham, and Northumberland*. Publications of the Surtees Society, vol. 97. Durham: Andrews and Company, 1897.

Palliser, David. "The Birth of York's Civic Liberties, c. 1200–1354." In *The Government of Medieval York*. York: Borthwick Studies in History, 3 (1997): 88–107.

———. *Medieval York 600–1540*. Oxford: Oxford University Press, 2014.

Panfilo, Gianna "The Coordinated Universal Time." *IEEE Instrumentation and Measurement Magazine* 19, no. 3 (2016): 28–33.

Panfilo, Gianna, Aurelie Harmegnies, and Laurent Tisserand. "A New Prediction Algorithm for the Generation of International Atomic Time." *Metrologia* 49, no. 1 (2012): 49–56.

———. "A New Weighting Procedure for UTC." *Metrologia* 51, no. 3 (2014): 285–92.

Panfilo, G, G. Petit, and A. Harmegnies. A First Step Towards the Introduction of Redundant Time Links for the Generation of UTC. *Metrologia* 57, no. 6 (2020). doi.org/10.1088/1681-7575/ab9d2e.

Pargman, Daniel, and Jacob Palme. "ASCII Imperialism." In *Standards and Their Stories*, edited by Martha Lampland and Susan Leigh Star, 177–99. Ithaca: Cornell University Press, 2009.

Parker, Thomas E., and Demetrios Matsakis. "Time and Frequency Dissemination: Advances in GPS Transfer Techniques." *GPS World* 15, November (2004): 32–38.

Parkinson, Bradford W., and Stephen T. Powers. "The Origins of GPS." *GPS World* 21, no. 5 (2010a): 30–41.

———. "Fighting to Survive." *GPS World* 21, no. 6 (2010b) 8–18.

Parrish, Jenni. "Litigating Time in America at the Turn of the Twentieth Century." *Akron Law Review* 36, no. 1 (2002): 1–47.

Pederson, Olaf. *A Survey of the Almagest*. New York: Springer, 2011.

Peil, Steven, James L. Hanssen, Thomas B. Swanson, Jennifer Taylor, and Christopher R. Ekstrom. "Evaluation of Long-Term Performance of Continuously Running Atomic Fountains." *Metrologia* 51, no 3. (2014): 263–69.

Peirce, Charles Sanders. "On a Method of Swinging Pendulums." *American Journal of Science* 18, issue 104 (1879): 112–19.

———. *Collected Papers of Charles Sanders Peirce, Volume 5.* Cambridge, MA: Belknap Press, 1,931–35.

———. *Collected Papers of Charles Sanders Peirce, Volume 7.* Cambridge, MA: Belknap Press, 1,931–35.

Pergman, Daniel. "Reflections on Cultural Bias and Adaptation." *Javnost—The Public* 6, no. 4 (1999): 23–38.

Peuerbach, Georg von. *Theoricae novae planetarum* [New Theories of the Planets]. Nuremberg: Regiomontanus, 1473.

Pickering, Andrew. *The Mangle of Practice: Time, Agency and Science.* Chicago: University of Chicago Press, 1995.

Podestà, Bartolomeo. "I Primi Oriuoli Pubblici in Bologna nei Secoli XIV e XV" [The first public clocks in Bologna in the fourteenth and fifteenth centuries]. *Atti e Memorie della R. Deputazione di Storia Patria per le Provincie di Romagna* [Proceedings and Memoirs of the Royal Deputation of National History for the Provinces of Romagna] 8 (1869): 141–65.

Polydore Vergil. *Urbinatis Anglicae Historiae* [The History of England]. Menston, Yorkshire: The Scolar Press, 1972.

Pope, Alexander. *An Essay on Criticism.* Menston, England: The Scolar Press, 1970.

Porter, Tony, and Liam Stockdale. "The Strategic Manipulation of Transnational Temporalities." *Globalizations* 13, no. 3 (2016): 270–84.

Postill, John. "Clock and Calendar Time: A Missing Anthropological Problem." *Time and Society* 11, no. 2/3 (2002): 251–70.

Postone, Moishe. *Time, Labor and Social Domination: A Reinterpretation of Marx's Critical Theory.* Cambridge: Cambridge University Press, 1993.

Potter, David. *The Bells and Bellringers of York Minster.* York: Friends of York Minster, 1987.

Press, Volker. "The Imperial Court of the Habsburgs: From Maximilian I to Ferdinand III, 1493–1657." In *Princes, Patronage, and the Nobility: The Court at the Beginning of the Modern Age c. 1450–1650*, edited by Ronald G. Asch and Adolf M. Birke, 289–312. Oxford: Oxford University Press, 1991.

Pschetz, Larissa. "Time Affordances and Physical Mobility in the Context of Ubiquitous Technologies." In *Human-Computer Interaction—Interact 2011*, edited by P. Campos, N. Graham, J. Jorge, N. Nunes, P. Palanque, and M. Winckler, 426–29. Heidelberg: Springer, 2011.

———. "Isn't It Time to Change the Way We Think About Time?" *Interactions* 22, no. 5 (2015): 58–61.

Pschetz, Larissa, and Michelle Bastian. "Temporal Design." *Design Studies* 56 (2017): 169–84.

Ptolemy, Claudius. *The Geography*, translated and edited by Edward Luther Stevenson. New York: Dover, 1991.

———. *Ptolemy's Almagest*, translated by G. J. Toomer. Princeton: Princeton University Press, 1998.

PTTI Panel Discussion. "Panel Discussion Following the Third DOD PTTI Strategic Planning Meeting." In *3rd Annual PTTI Meeting*, 395–415. Washington, DC: United States Naval Observatory, 1971.

Putkovich, Kenneth. "USNO GPS Program." *Proceedings of the Precise Time and Time Interval (PTTI) Systems and Applications Meeting.* 12 (1980): 387–415.

Quiccheberg, Samuel. *Inscriptiones Vel Tituli Theatri Amplissimi* [Inscriptions or Titles of the Most Magnificent Theater]. Munich: Adam Berg, 1565.

Quinn, Terry. "The BIPM and the Accurate Measurement of Time." *Proceedings of the IEEE* 79, no. 7 (1991): 894–905.

———. "Time, the SI and the Metre Convention." *Metrologia* 48, no. 4 (2011): S121–24.

———. *From Artefacts to Atoms: The BIPM and the Search for Ultimate Measurement Standards.* Oxford: Oxford University Press, 2012.

Rabanus Maurus. "De Computo" [About the Computus]. In *Patrologia Latina, volume 107*, edited by J. –P. Migne, columns 669–728. Turnhold: Typographi Brepols Editores Pontifici, 1844–1864.

Rabkin, Yakov M. "Rediscovering the Instrument: Research, Industry, and Education." In *Invisible Connections: Instruments, Institutions, and Science*, edited by Robert Bud and Susan E. Cozzens, 57–72. Bellingham, WA: SPIE Optical Engineering Press, 1992.

Raine, Angelo, ed. *York Civic Records, volume III.* Record Series volume CVI. Wakefield: Yorkshire Archaeological Society, 1942.

———. *York Civic Records, volume V.* Record Series volume CX. Wakefield: Yorkshire Archaeological Society, 1946.

Ramelli, Agostino. *Le diverse et artificiose machine* [The various and ingenious machines]. Paris: Agostino Ramelli, 1588.

Ramus, Petrus. *Scholarum mathematicarum libri unus et triginta* [Thirty-one books of the mathematical schools]. Basil: Eusebius Episcopius, 1569.

Ravipati, Sri. "The Moon Could Get Its Own Timezone." *Axios* 27 February 2023. https://www.axios.com/2023/02/28/moon-timezone?utm_campaign. Retrieved 5 January 2026.

Rees Jones, Sarah. "York's Civic Administration, 1354–1464." In *The Government of Medieval York*, edited by Sarah Rees Jones, 108–40. York: Borthwick Studies in History, 3, 1997.

———. 2013. *York: The Making of a City 1068–1350.* Oxford: Oxford University Press, 2013.

Rees Jones, Sarah, and D. M. Palliser. "York, 1271–1536." In *York: British Historic Towns Atlas, volume V*, edited by Peter Addyman, 37–48. Oxford: Oxbow Books, 2016.

Regiomontanus, Johannes. *Ephemerides*. Nuremberg: Regiomontanus, 1474.

———. *Kalendarium* [Calendar] Nuremberg: Regiomontanus, 1474.

———. "Observationes Ioannes de Monteregio per regulas Ptolomaei factae ad Solem" [Observations of Regiomontanus according to the Sun by Ptolemy's Rules]. Appended to *Coeli & siderum in eo errantium Observationes Hassiacae* [Observations of the Sky and Wandering Stars in Hesse], edited by Willibord Snellius. Leiden: Lugdunum Batavorum, 1618.

———. "Joannes de regiomonte magistro Christiano [Roder] mathematicorum prestantissimo." [Regiomontanus to the Outstanding Mathemetician Christian Roder]. In *Memorabilia bibliothecarum publicarum Norimbergensium et Universitatis Altdorfinae* [Memorabilia of the Public Libraries of Nuremberg and the University of Altdorf], edited by Christoph Gottlieb von Murr, 184–205. Nuremberg: Hoesch, 1786.

Regiomontanus, Johannes, and Johannes Schöner. *Scripta Clarissimi Mathmatici* [Writings of the Famous Mathemetician]. Nuremberg: Ioannes Montanus and Ulric Neuber, 1544.

Reinhardt, Victor. "Relativistic Effects of the Rotation of the Earth on Remote Clock Synchronization." *Proceedings of the Precise Time and Time Interval (PTTI) Systems and Applications Meeting* 6 (1974): 395–424.

Reindl-Kiel, Hedda. "Ottoman-European Cultural Exchange: East is East and West is West, and Sometimes the Twain Did Meet Diplomatic Gift Exchange in the Ottoman Empire." *Frontiers of Ottoman Studies: State, Province, and the West*, 2 (2005): 113–23.

R. F. *Kalender.* Germanishe Nationalmuseum. Inventory HB 2675, 1564. Retrieved 5 January 2026 from https://objektkatalog.gnm.de/wisski/navigate/23861/view.

Rice, Prudence M. *Maya Political Science: Time, Astronomy, and the Cosmos.* Austin: University of Texas Press, 2004.

———. *Maya Calendar Origins: Monuments, Mythistory, and the Materialization of Time.* Austin: University of Texas Press, 2007.

Rich, Burdett A., and Henry P. Farnham, eds. *The Lawyers Reports Annotated*. Rochester, NY: The Lawyers Cooperative Publishing Company, 1912.

Richards, J. D. *The Bedern Foundry*. The Archaeology of York, vol. 10, fascicule 3. York: York Archaeological Trust, 1993.

Richardson, Gary. "Guilds, Laws, and Markets for Manufactured Merchandise in Late-Medieval England." *Explorations in Economic History* 41, no. 1 (2004): 1–25.

Riddick, Floyd M., and Miriam H. Butcher. *Riddick's Rules of Procedure*. New York: Charles Scribner's Sons, 1985.

Rimple, Mark T. "The Enduring Legacy of Boethian Harmony." In *A Companion to Boethius in the Middle Ages*, edited by Noel Harold Kaylor, Jr. and Philip Edward Phillips, 447–78. Lieden: Brill, 2012.

Robertson, J. Drummond. *The Evolution of Clockwork*. East Ardsley, Yorkshire: EP Publishing, 1982.

Rocca, Angelo. *De Campanis Commentarius* [A Commentary on Bells]. Rome: Guglielmo Facciotti, 1612.

Rodríguez-Salgado, M. J. "The Court of Philip II of Spain." In *Princes, Patronage, and the Nobility: The Court at the Beginning of the Modern Age c. 1450–1650*, edited by Ronald G. Asch and Adolf M. Birke, 205–44. Oxford: Oxford University Press, 1991.

Rogers, Nicola, and Richard Hall. "A Striking Discovery: The Seal of a Horologiarius." *Antiquarian Horology* 30, no. 2 (2007): 223–24.

Rooney, David. *About Time*. New York: W. W. Norton, 2021.

Rooney, David, and James Nye. "'Greenwich Observatory Time for the Public Benefit': Standard Time and Victorian Networks of Regulation." *British Society for the History of Science* 42, no. 1 (2009): 5–30.

Rosa, H. *Social Acceleration: A New Theory of Modernity*. New York: Columbia University Press, 2013.

Rose, Paul Lawrence. *The Italian Renaissance of Mathematics: Studies on Humanists and Mathematicians from Petrarch to Galileo*. Geneva: Librairie Droz, 1975.

Rossi, Paolo. *Philosophy, Technology and the Arts in the Early Modern Era*, translated by Salvator Attanasio. New York: Harper and Row, 1970.

Rothwell, W. "The Hours of the Day in Medieval French," *French Studies* 13, no. 3 (1959): 240–51.

Rotter, George S. "Clock-speed as an Independent Variable in Psychological Research." *The Journal of General Psychology* 81 (1): 45–52.

Royal Commission on Historical Monuments. *An Inventory of the Historical Monuments in City of York, Volume 2, the Defenses*. London: Her Majesty's Stationary Office, 1972. Retrieved 5 January 2026 from https://www.british-history.ac.uk/rchme/york/vol2.

Rubbiani, Alfonso. "L'Orologio del commune di Bologna e la Sfera Del 1451" [Bologna's City Clock and the Sphere of 1451]. *Atti e Memorie della R. Deputazione di Storia Patria per le Provincie di Romagna* [Proceedings and Memoirs of the Royal Deputation of National History for the Provinces of Romagna], 3rd ser., 26 (1908): 349–66.

Sacrobosco. The *Sphere* of Jannes de Sacrobosco. In *The Sphere of Sacrobosco and Its Commentators*, edited and translated by Lynn Thorndike, 76–142. Chicago: University of Chicago Press, 1949.

Sadler, D. H. "Ephemeris Time." *Occasional Notes, Royal Astronomical Society* 17, no. 3 (1954): 103–13.

———. *Transactions of the International Astronomical Union, volume XIA: Reports on Astronomy*. London: Academic Press, 1962.

———. "The New System of Coordinated Universal Time." *Journal of Navigation* 25, no. 1 (1972): 32–42.

———. "A Personal History of the H.M. Nautical Almanac Office," edited by George A. Wilkins. Taunton: United Kingdom Hydrographic Office, 2008.

Sassen, Saskia. "Spatialities and Temporalities of the Global." *Public Culture* 12, no. 1 (2000): 215–32.

———. "The Global City: Introducing a Concept." *Brown Journal of World Affairs* 11, no. 2 (2005): 27–43.

Saussure, Ferdinand de. *Course in General Linguistics*, edited by Charles Bally, Albert Sechehaye, and Albert Riedlinter, translated by Wade Baskin. New York: McGraw Hill, 1966.

Sauter, Michael J. "Clockwatchers and Stargazers." *American Historical Review* 112, no. 3 (2007): 685–709.

Sayer, Peter. "8 More Years of Leap-Second Problems Loom; Governments Punt Decision to 2023." *PC World*, 20 November 2015. Retrieved 16 November 2015 from https://www.pcworld.com/article/424487/eight-more-years-of-leap-second-problems-loom-as-governments-punt-decision-to-2023.html.

Schaefer, Bradley. "Sherlock Holmes and Some Astronomical Connections." *Journal of the British Astronomical Association* 103, no. 1 (1993): 30–34.

Scheibe, A., and U. Adelsberger. "Nachweis von Schwankungen der astronomischen Tageslänge und der astronomischen Zeitbestimmung mit den Quarzhurhen der Physikalisch-Technischen Reichsanstalt" [Detection of Fluctuations in the Astronomical Day Length and Astronomical Time Determination by the Quartz Clocks of the Physikalisch-Technische Reichsanstalt]. *Physikalische Zeitschrift* [Physical Journal] 37 (1936): 185–203.

Scheicher, Elisabeth. *Die Kunst- und Wunderkammern der Habsberger* [The Habsburg Cabinets of Curiosities]. Vienna: Molden, 1979.

Scheuerman, W. *Liberal Democracy and the Social Acceleration of Time*. Baltimore: Johns Hopkins University Press, 2004.

Schillinger, Klaus. "The Development of Saxon Scientific Instrument-Making Skills from the Sixteenth Century to the Thirty Years War." *Annals of Science* 47, no. 3 (1990): 277–89.

Schlosser, Julius von. *Die Kunst- und Wunderkammern der Spätrenaissance* [The Cabinets of Curiosities of the Late Renaissance]. Leipzig: Verlag von Klinkhardt & Biermann, 1908.

Schmidgen, Henning. "Brain-Time Experiments." In *Power and Time*, edited by Dan Edelstein, Stefanos Geroulanos, and Natasha Wheatley, 224–248. Chicago: University of Chicago Press, 2020.

Schneider, R. J. "Time and Frequency Standardization." *Proceedings of the Precise Time and Time Interval (PTTI) Systems and Applications Meeting*. 2 (1971): 7–21.

Schöner, Johannes. *Globi stelliferi, sive sphaerae stellarum fixarum usus* [The Use of the Celestial Globe or Sphere of Fixed Stars]. Nuremberg: Johannes Petreium, 1533.

———. *Opusculum Astrologicum e diversorum libris summa cura pro studiosorum utilitate collectum* [A Little Astrological Work Compiled with Great Care from Various Books for the Use of Students]. Nuremberg: Johannes Petreium, 1539.

Schöner, Andreas. *Gnomonice* [Gnomonics]. Nuremberg: Montanus and Neuberus, 1562.

Schöner, Andreas, and Wilhelm IV. "Beobachtungen und Aufzeichnungen zu den Indikationen des Uhrwerks der großen Planetenuhr" [Observations and Records Indicating the Movements for a Great Planetary Clock]. Bound with Apian 1540, 4–72 and 192–245, n.d. Retrieved 29 October 2019 from https://orka.bibliothek.uni-kassel.de/viewer/thumbs/1378193947325/9/.

Schukowski, Manfred. *Wunderuhren: Astronomische Uhren in Kirchen der Hansezeit* [Marvelous Clocks: Astronomical Clocks in Churches of the Hanseatic League]. Schwerin: Thomas Helms Verlag, 2006.

Scott, David, and Mike Cowham. *Time Reckoning in the Medieval World: A Study of Anglo-Saxon and Early Norman Sundials*. British Sundial Society Monograph, no. 8, 2010.

Scott, James. *Seeing Like a State*. New Haven: Yale University Press, 1998.

Seago, J., and M. F. Storz. "UTC Redefinition and Space and Satellite-Tracking Systems." Proceedings of the ITU-R SRG Colloquium on the UTC Timescale IEN Galileo Ferraris, Torino, Italy, 2003.

Seago, John H., Robert L. Seaman, and Steven Allen, eds. *Decoupling Civil Timekeeping from Earth Rotation*. American Astronautical Society, Science and Technology Series, vol. 113. San Diego: Univelt, 2011.

Seago J. H., R. L. Seaman, and P. K. Seidelman. "The Colloquium on Requirements for UTC and Civil Timekeeping on Earth." In *Requirements for UTC and Civil Timekeeping on Earth*, edited by J. H. Seago, R. Seaman, P. K. Seidelmann, and S. L. Allen, 3–12. American Astronautical Society, Science and Technology Series, vol. 115. San Diego: Univelt, 2013.

Seago, J. H., R. L. Seaman, P. K. Seidelmann, and S. L. Allen, eds. *Requirements for UTC and Civil Timekeeping on Earth*. American Astronautical Society, Science and Technology Series, vol. 115. San Diego: Univelt, 2013.

Seaman, Robert. Systems Engineering for Civil Timekeeping. In *Decoupling Civil Timekeeping from Earth Rotation*, edited by John H. Seago, Robert L. Seaman, and Steven Allen. American Astronautical Society, Science and Technology Series, vol. 113, 15–25. San Diego: Univelt, 2011.

Seaman, Robert, Frank Shelly, Eric Christensen, Alexander Gibbs, and Stephen Larson. "Keeping Time with the Asteroids." In *The Science of Time 2016: Time in Astronomy and Society, Past, Present and Future*, edited by Elisa Felicitas Arias, Ludwig Combrinck, Pavel Gabor, Catherine Hohenkerk, and P. Kenneth Seidelmann, 329–30. Cham, Switzerland: Springer, 2017.

Seed, Patricia. *Ceremonies of Possession in Europe's Conquest of the New World, 1492–1640*. Cambridge: Cambridge University Press, 1995.

Seidelmann, P. Kenneth, and John H. Seago, 2011. "Time Scales, Their Users, and Leap Seconds." *Metrologia* 48, no. 4 (2011): S186–94.

Sellers, Maud. *The York Memorandum Book, Part I, 1376–1419*. Surtees Society, vol. 120. Durham: Andrews and Company, 1912.

Serres, Michel, and Bruno Latour. *Conversations on Science, Culture, and Time*, translated by Roxanne Lapidus. Ann Arbor: University of Michigan Press. 1995.

Shapin, Steven, and Simon Schaffer. *Leviathan and the Air-Pump: Hobbes, Boyle, and the Experimental Life*. Princeton: Princeton University Press, 1985.

Sharma, Sarah. *In the Meantime: Temporality and Cultural Politics*. Durham, NC: Duke University Press, 2014.

Siddall, M, E. J. Rohling, A. Almogi-Labin, Ch. Hemleben, D. Meischner, I. Schmetzer, and D. A. Smeed. "Sea-level Fluctuations during the Last Glacial Cycle." *Nature* 423 (2003): 853–58.

Sivin, Nathan. *Granting the Seasons: The Chinese Astronomical Reform of 1280, with a Study of Its Many Dimensions and an Annotated Translation of Its Records*. New York: Springer, 2009.

Smith, Adam. *Wealth of Nations*. New York: The Modern Library.

Smith, Humphry Montague. "International Time and Frequency Coordination." *Proceedings of the IEEE* 60, no. 5 (1972): 479–87.

Smith, William. "A Description of the Cittie of Noremberg" (Beschreibung der Reichstadt Nürnberg) 1594. *Mitteilungen des Vereins für Geschichte der Stadt Nürnberg* [Reports of the Association for the History of the City of Nuremberg] 48 (1958): 194–245.

Sôma, Mitsuru, Kin-aki Kawabata, and Kiyotaka Tanikawa. "Units of Time in Ancient China and Japan." *Publications of the Astronomical Society of Japan* 56, no. 5 (2004): 887–904.

Spivak, Gayatri Chakravorty. "Can the Subaltern Speak?" In *Colonial Discourse and Post-Colonial Theory*, edited by Patrick Williams and Laura Chrisman, 66–111. New York: Columbia University Press, 1988.

Star, Susan Leigh. 1991. "Power, Technologies and the Phenomenology of Conventions: On being Allergic to Onions." In *A Sociology of Monsters: Essays on Power, Technology and Domination*, edited by John Law, 26–56. London: Routledge, 1991.

Star, Susan Leigh, and Martha Lampland. "Reckoning with Standards." In *Standards and Their Stories*, edited by Martha Lampland and Susan Leigh Star, 3–24. Ithaca: Cornell University Press, 2009.

Staudacher, Fritz. *Jost Bürgi, Kepler und der Kaiser: Uhrmacher, Instrumentbauer, Astronom, Mathematiker, 1552–1632* [Jost Bürgi, Kepler and the Kaiser: Clockmaker, Instrument builder, Astronomer, Mathematician, 1552–1632]. Zürich: Verlag Neue Zürcher Zeitung, 2013.

Stephens, Carlene E. *On Time: How America Has Learned to Live by the Clock.* Boston: Little, Brown and Company, 2002.

Stern, Sacha. *Calendars in Antiquity*. Oxford: Oxford University Press, 2012.

Stevens, Wesley. *Cycles of Time and Scientific Learning in Medieval Europe*. Aldershot: Variorum 1995.

Stiegler, Bernard. *Technics and Time, 1: The Fault of Epimetheus*, translated by Richard Beardsworth and George Collins. Stanford, CA: Stanford University Press, 1998.

———. *Technics and Time, 2: Disorientation*, translated by Stephen Barker. Stanford, CA: Stanford University Press, 2009.

———. *Technics and Time, 3: Cinematic Time and the Question of Malaise*, translated by Stephen Barker. Stanford, CA: Stanford University Press, 2011.

———. *What Makes Life Worth Living*, translated by Daniel Ross. Cambridge: Polity, 2013.

———. *Automatic Society*, translated by Daniel Ross. Cambridge: Polity Press 2016.

———. *The Age of Disruption*, translated by Daniel Ross. Cambridge: Polity, 2019.

———. For a Neganthropology of Automatic Society. In *Machine*, 25–47. Minneapolis: University of Minnesota Press, 2019.

Stoyko, M. Nicolas. "Sur l'irrégularité de la rotation de la Terre" [On the Irregularity of Earth's Rotation]. *Comptes rendus hebdomadaires des séances de l'Académie des Sciences* [Weekly Reports of the Meetings of the Academy of Sciences] 203 (1936): 39–40.

———. "Sur la périodicité dans l'irrégularité de la rotation de la Terre." *Comptes rendus hebdomadaires des séances de l'Académie des Sciences* [Weekly Reports of the Meetings of the Academy of Sciences] 205 (1937): 79–81.

Strathern, Marilyn. *Reproducing the Future: Essays on Anthropology, Kinship, and the New Reproductive Technologies*. New York: Routledge, 1992.

Stromer, Wolfgang von. "Hec Opera Fient in Oppido Nuremerga Germanie Ductu Ioannis de Monteregio" [These works will be done in Nuremberg, Germany, under the leadership of Regiomontanus]. In *Regiomontanus-Studien*, edited by Günther Hamann, 267–89. Vienna: Verlag der Österreichischen Akademie der Wissenschaften, 1980.

Swanson, Heather. "The Illusion of Economic Structure: Craft Guilds in Late Medieval English Towns." *Past and Present* 121 (1988): 29–48.

———. *Medieval Artisans: An Urban Class in Late Medieval England*. Oxford: Basil Blackwell, 1989.

Swinehart, Karl. "Decolonial Time in Bolivia's *Pachakuti*." *Signs and Society* 7, no. 1 (2019): 96–114.

Swinford, S. 2014. "Greenwich Mean Time Could Drift to the US, Minister Warns." *Daily Telegraph*, 14 May 2014. Retrieved 30 September 2015 from http://www.telegraph.co.uk/news/politics/10831974/Greenwich-Mean-Time-could-drift-to-the-US-minister-warns.html.

Syracuse Daily Courier. "New York Central Railroad." 28 April 1858: 4.

Tavella, Patrizia. "Coordinated Universal Time: An Overview." *ITU News Magazine—The Future of Coordinated Universal Time*, no. 2 (2023): 15–18.

Taylor, Charles. *A Secular Age*. Cambridge, MA: Belknap, 2007.

Taylor, Frederic Winslow. 1967. *Principles of Scientific Management*. New York: Norton, 1967.

ten Hoeve, Johan, and David Thompson. "A Flemish Clock at the Shogun's Shrine." *Antiquarian Horology* 35, no. 4 (2014): 1,063–76.

Thévenot, Laurent. "Governing Life by Standards." *Social Studies of Science* 39, no. 5 (2009): 793–813.

———. "Measure for Measure." *Historical Social Research* 44, no. 2 (2019): 44–76.

Thomas, Claudine, Peter Wolf, and Patrizia Tavella. "Time Scales." *Monographie*, no. 1. Sèvres: BIPM, 1994.

Thomas, Keith. *Religion and the Decline of Magic*. New York: Scribner, 1971.

Thomas, Nicholas. *Entangled Objects: Exchange, Material Culture, and Colonialism in the Pacific*. Cambridge, MA: Harvard University Press, 1991.

Thompson, E. P. "Time, Work-discipline, and Industrial Capitalism." *Past and Present* 38, no. 1 (1967): 56–97.

Thönes, Sven, Stefan Arnau, and Edmund Wascher. "Cognitions about Time Affect Perception, Behavior, and Physiology." *Consciousness and Cognition* 63 (2018): 99–109.

Thorndike, Lynn. "The True Place of Astrology in the History of Science." *Isis* 46, no. 3 (1955): 273–78.

Thrift, Nigel. "An Episode in the Development of Capitalist Time Consciousness: A Case Study of the Diffusion of Greenwich Mean Time." In *The Sociology of Time*, edited by John Hassard, 120–29. New York: St. Martin's Press, 1990.

Thrupp, S. *The Merchant Class of Medieval London*. Ann Arbor: University of Michigan Press, 1948.

Tolhurst, J. B. L., ed. *The Ordinal and Customary of the Abbey of Saint Mary, York.* Henry Bradshaw Society, vol. LXXII. Rochester, NY: The Boydell Press, 2010.

Torre, Antonii van. *Dialogi Familiares Litterarum Tironibus* [Familiar Dialogies of Literature for Beginners]. Antwerp: Willerval, 1746.

Trouillot, Michel-Rolph. *Silencing the Past: Power and the Production of History.* Boston: Beacon, 1995.

Tullis, Paul. "Screwed in a Millisecond." *Bloomberg Businessweek*, 30 July 2018: 50–52.

Turnbull, David. "Cartography and Science in Early Modern Europe: Mapping the Construction of Knowledge Spaces." *Imago Mundi* 48, no. 1 (1996): 5–24.

Turner, Anthony. *Early Scientific Instruments: Europe 1400–1800.* London: Sotheby's Publications, 1987.

Turner, Gerard l'E. "The Cabinet of Experimental Philosophy." In *The Origins of Museums: The Cabinet of Curiosities in Sixteenth- and Seventeenth-Century Europe*, edited by Oliver Impey and Arthur MacGregor, 295–307. Thirsk, North Yorkshire: House of Stratus, 2001.

United States General Accounting Office. *Patriot Missile Defense Software Problem Led to System Failure at Dhahran, Saudi Arabia.* Washington, DC: US GAO, 1992.

Uribe, Ricardo. *Las Dinámicas del Tiempo: Relojes, Calendarios y Actitudes en el Virreinato de la Nueva Granada.* Medellin: La Carreta Editores, 2016.

Vehse, E. *Memoirs of the Court, Aristocracy, and Diplomacy of Austria, volume 1*, translated by Franz Demmler. London: Longman, Brown, Green, and Longmans, 1856.

Vázquez, Rolando. 2009. "Modernity, Coloniality, and Visibility." *Sociological Research Online* 14 (2009). https://doi.org/10.5153/sro.1990.

Velkar, Aashish. "Rethinking Metrology, Nationalism and Development in India, 1833–1956." *Past and Present* 239 (2018): 143–79.

Veranzio, Fausto. *Machinae Novae* [New Machines]. Venice: n.p., 1617.

Viani, Bonaventura. *Elementi di cronologia utili e dilettevoli* [Useful and enjoyable elements of chronology]. Oneglia: Giovanni Ghilini, 1878.

Vico, Giambattista. *The New Science*, translated by Thomas Goddard Bergin and Max Harold Fisch. Ithaca: Cornell University Press, 1968.

Vilhjalmsson, Thorsteinn. "Time and Travel in Old Norse Society." *Disputatio* 2 (1997): 89–114.

Virilio, Paul. *The Lost Dimension*, translated by Daniel Moshenberg. New York: Semiotext(e), 1991.

———. *A Landscape of Events*, translated by Julie Rose. Cambridge, MA: MIT Press, 2000.

———. *Speed and Politics, new edition*, translated by Mark Polizzotti. Cambridge, MA: MIT Press, 2006.

———. *The Great Accelerator*, translated by Julie Rose. Cambridge: Polity, 2012.

Virilio, Paul, and Sylvere Lotringer. *Crepescular Dawn*, translated by Mike Taormina. Los Angeles: Semiotext(e), 2002.

Vitruvius. *Vitruvius: The Ten Books on Architecture*, translated by Morris Hicky Morgan. New York: Dover, 1960.

Vygotsky, Lev S. *The Collected Works of L. S. Vygotsky, Volume 1: Problems of General Psychology Including the Volume Thinking and Speech*, edited by Robert W. Rieber and Aaron S. Carton, translated by Norris Minick. New York: Plenum Press, 1987.

———. *The Collected Works of L. S. Vygotsky, Volume 3: Problems of the Theory and History of Psychology*, edited by Robert W. Reiber and Jeffrey Wollock, translated by René van der Veer. New York: Plenum Press, 1997.

Wade, Nicholas J., and Josef Brožek. *Vision: The Dawning of Neuroscience.* Mahwah, NJ: Lawrence Erlbaum, 2001.

Wajcman, Judy. "Life in the Fast Lane? Towards a Sociology of Technology and Time." *The British Journal of Sociology* 59, no. 1 (2008): 59–77.

Wallace, P. T. Software for Timescale Applications. *Metrologia* 48, no. 4 (2011): S200–2.

Wallis, Helen M., and Arthur H. Robinson, eds. *Cartographical Innovations*. Tring, UK: Map Collector Publications and International Cartographic Association, 1987.

Walls, Hoy J. "The Standard-Frequency Set at WWV." *QST Magazine*, October (1924): 9–12.

Weber, Max. *Economy and Society*, edited and translated by Guehnter Roth and Claus Wittich. Berkeley: University of California Press, 1978.

Wellington Journal. "Selling Ale During Prohibited Hours." 1 January 1876: 6.

Wells, Cheryl. *Civil War Time*. Athens: University of Georgia Press, 2005.

Werner, Theodor. "Nürnbergs Erzeugung und Ausfuhr wissenschaftlicher Geräte im Zeitalter der Entdeckungen" [Nuremberg's production and exporting of scientific equipment in the age of discover]. *Mitteilungen des Vereins für Geschichte der Stadt Nürnberg* [Reports of the Association for the History of the City of Nuremberg] 53 (1965): 69–149.

West-Pavlov, Russell. *Temporalities*. London: Routledge, 2013.

Western Daily Press. "The New Licensing Act." 1 October 1872: 3.

———. "Enforcement of the Licensing Act at Keynsham." 21 December 1872: 3.

Western Times. "Appeal Court: James Westaway Appellant." 16 October 1873: 3.

Whibberley, Peter B., John A. Davis, and Setnam L. Shemar. "Local Representations of UTC in National Laboratories. "*Metrologia* 48, no. 4 (2011): S154–64.

White, Lynn. 1962. *Medieval Technology and Social Change*. Oxford: Oxford University Press, 1962.

———. *Medieval Religion and Technology*. Berkeley: University of California Press, 1978.

Whitehead, Alfred North. *An Introduction to Mathematics*. London: Williams and Northgate, 1911.

———. *Science and the Modern World*. New York: New American Library, 1963.

Whitrow, G. J. *Time in History: Views of Time from Prehistory to the Present Day*. Oxford: Oxford University Press, 1988.

Widdington, Thomas. *Analecta Eboracensia: Some Remaynes of the Ancient City of York*. London: C. J. Clark, 1897.

Wilkin, Alexis. "Time Constraints on Market Activity and the Balance of Power in Medieval Liège." *Continuity and Change* 30, no. 3 (2015): 315–40.

Wilson, Barbara, and Frances Mee. *'The Fairest Arch in England': Old Ouse Bridge, York, and its Buildings*. York: York Archaeological Trust, 2002.

Winkler, G. M. R. "Summary of Comments on Present and Future Time Synchronization Methods." *Report on Department of Defense Precise Time and Time Interval (PTTI) Conference*, Attachment 2, 1–3. 1969.

———. "UTC Worldwide Adjustment." *Proceedings of the Precise Time and Time Interval (PTTI) Systems and Applications Meeting* 2 (1971): 387–93.

———. "Developments in the Time Service Division." In *Sky with Ocean Joined: Proceedings of the Sesquicentennial Symposia, U. S. Naval Observatory*, edited by Steven J. Dick and LeRoy E. Doggett, 115–31. Washington, DC: U.S. Naval Observatory, 1983.

Winner, Langdon. "Do Artifacts Have Politics?" *Daedalus* 109, no. 1 (1980): 121–36.

Wishnitzer, Avna. *Reading Clocks, Alla Turca: Time and Society in the Late Ottoman Empire*. Chicago: University of Chicago Press, 2015.

———. "At Approximately Eleven, Just Before Nightfall": An Introduction to Ottoman Temporal Culture." In *Society, Law and Culture in the Middle East*, edited by Dror Ze'evi, Ehud R. Toledano, and Dror Ze'evi, 121–134. Warsaw: De Gruyter Open, 2015.

Wöckel, Lorenz. *Populäre Vorlesungen* über *die Sternkunde* [Popular Lectures in Astronomy]. Nuremberg: Bauer und Raspe, 1844.

Wolfschmidt, Gudrun. "The Observatories and Instruments of Tycho Brahe." In *Tycho Brahe and Prague: Crossroads of European Science*, edited by J. R. Christianson, Alena Hadravová, Petr Hadrava, and Martin Šolc, 203–16. Frankfurt am Main: Verlag Harri Deutsch, 2002.

WRC-12. "Resolution 653: Future of the Coordinated Universal Time Time-scale." Geneva: ITU, 2012.

WRC-23. "Resolution 655: Definition of time scale and dissemination of time signals via radiocommunication systems." Geneva: ITU, 2023.

Wrexham Advertiser. "Important to Publicans." 28 November 1874: 3.

Wu Hung. "The Hong Kong Clock—Public Time-Telling and Political Time/Space." *Public Culture* 9, no. 3 (1997): 329–54.

Yokota, Yasuhiro. A Historical Overview of Japanese Clocks and Karakuri. In *International Symposium on History of Machines and Mechanisms*, edited by Yong-Sen Yan and Marco Ceccarelli, 175–88. Dordrecht: Springer, 2008.

York Herald. "What is Local Time for Legal Purposes?" 18 May 1880: 4.

York Minster. *The Fabric Rolls of York Minster*. Durham: Publications of the Surtees Society, vol. 35, 1859.

Zallinger zum Thurn, Jacobus Antonius Zallinger. *Institutionum juris naturalis et ecclesiastic publici, libri V* [Institutions of natural and ecclesiastical law, book V]. Collegium Urbanum, 1832.

Ziegler, Alexander. *Regiomontanus—Joh. Müller aus* Königsberg in Franken—*ein geistiger Vorläufer des Columbus* [Joh. Müller from Königsberg in Franconia—a spiritual predecessor of Columbus]. Amsterdam: Meridian Publishing Company, 1967.

Ziggelaar, August. "The Papal Bull of 1582 Promulgating a Reform of the Calendar. In *Gregorian Reform of the Calendar*," edited by G. V. Coyne, M. A. Hoskin, and O. Pederson, 201–39. Vatican City: Specola Vaticana, 1983.

Zika, Charles. "Hosts, Processions and Pilgrimages: Controlling the Sacred in Fifteenth-Century Germany." *Past & Present* 118, no. 1 (1988): 25–64.

Zinner, Ernst. *Regiomontanus: His Life and His Work*, translated by Ezra Brown. Amsterdam: North Holland, 1990.

INDEX

www.ingramcontent.com/pod-product-compliance
Lightning Source LLC
LaVergne TN
LVHW011813200926
848484LV00005B/1174